THE STRANGE ORDER OF THINGS

THE STRANGE ORDER
of THINGS

Life, Feeling, and the Making of Cultures

ANTONIO DAMASIO

PANTHEON BOOKS, NEW YORK

Published in the United States by Pantheon Books, a division
of Penguin Random House LLC, New York, and distributed
in Canada by Random House of Canada, a division of
Penguin Random House Canada Limited, Toronto.

Pantheon Books and colophon are registered trademarks
of Penguin Random House LLC.

Library of Congress Cataloging-in-Publication Data
Names: Damasio, Antonio R., author.
Title: The strange order of things : life, feeling, and the making
of cultures / Antonio Damasio.
Description: New York : Pantheon Books, 2018. Includes
bibliographical references and index.
Identifiers: LCCN 2017019925. ISBN 9780307908759
(hardcover : alk. paper). ISBN 9780307908766 (ebook)
Subjects: LCSH: Homeostasis. Neurosciences.
Classification: LCC QP90.4 .D36 2017 | DDC 612/.022—dc23 |
LC record available at lccn.loc.gov/2017019925

www.pantheonbooks.com

Jacket design by Kelly Blair

Printed in the United States of America
First Edition
9 8 7 6 5 4 3 2 1

For Hanna

I see it feelingly.

—Gloucester to Lear
SHAKESPEARE, *King Lear,* act 4, scene 6

The fruit is blind. It's the tree that sees.

—RENÉ CHAR

CONTENTS

PART II   ASSEMBLING THE CULTURAL MIND

THE STRANGE ORDER OF THINGS

BEGINNINGS

1

This book is about one interest and one idea. I have long been intrigued in human affect—the world of emotions and feelings—and have spent many years investigating it: why and how we emote, feel, use feelings to construct our selves; how feelings assist or undermine our best intentions; why and how brains interact with the body to support such functions. I have new facts and interpretations to share on these matters.

As for the idea, it is very simple: feelings have not been given the credit they deserve as motives, monitors, and negotiators of human cultural endeavors. Humans have distinguished themselves from all other beings by creating a spectacular collection of objects, practices, and ideas, collectively known as cultures. The collection includes the arts, philosophical inquiry, moral systems and religious beliefs, justice, governance, economic institutions, and technology and science. Why and how did this process begin? A frequent answer to this question invokes an important faculty of the human mind—verbal

language—along with distinctive features such as intense sociality and superior intellect. For those who are biologically inclined the
answer also includes natural selection operating at the level of genes.
I have no doubt that intellect, sociality, and language have played key
roles in the process, and it goes without saying that the organisms
capable of cultural invention, along with the specific faculties used in
the invention, are present in humans by the grace of natural selection
and genetic transmission. The idea is that something else was required
to jump-start the saga of human cultures. That something else was a
motive. I am referring specifically to feelings, from pain and suffering
to well-being and pleasure.

Consider medicine, one of our most significant cultural enterprises.
Medicine's combination of technology and science began as a response
to the pain and suffering caused by diseases of every sort, from physical trauma and infections to cancers, contrasted with the very opposite
of pain and suffering: well-being, pleasures, the prospect of thriving.
Medicine did not begin as an intellectual sport meant to exercise one's
wits over a diagnostic puzzle or a physiological mystery. It began as
a consequence of specific feelings of patients and specific feelings of
early physicians, including but not limited to the compassion that
may be born of empathy. Those motives remain today. No reader will
have failed to notice how visits to the dentist and surgical procedures
have changed for the better in our own lifetime. The primary motive
behind improvements such as efficient anesthetics and precise instru-
mentation is the management of feelings of discomfort. The activity
of engineers and scientists plays a commendable role in this endeavor,
but it is a motivated role. The profit motive of the drug and instrumen-
tation industries also plays a significant part because the public does
need to reduce its suffering and industries respond to that need. The
pursuit of profit is fueled by varied yearnings, a desire for advance-
ment, prestige, even greed, which are none other than feelings. It is
not possible to comprehend the intense effort to develop cures for cancers or Alzheimer's disease without considering feelings as motives,

monitors, and negotiators of the process. Nor is it possible to comprehend, for example, the less intense effort with which Western cultures have pursued cures for malaria in Africa or the management of drug addictions most everywhere without considering the respective web of motivating and inhibiting feelings. Language, sociality, knowledge, and reason are the primary inventors and executors of these complicated processes. But feelings get to motivate them, stay on to check the results, and help negotiate the necessary adjustments.

The idea, in essence, is that cultural activity began and remains deeply embedded in feeling. The favorable *and* unfavorable interplay of feeling and reason must be acknowledged if we are to understand the conflicts and contradictions of the human condition.

2

How did humans come to be at the same time sufferers, mendicants, celebrants of joy, philanthropists, artists and scientists, saints and criminals, benevolent masters of the earth and monsters intent on destroying it? The answer to this question requires the contributions of historians and sociologists, for certain, as well as those of artists, whose sensibilities often intuit the hidden patterns of the human drama, but the answer also requires the contributions of different branches of biology.

As I considered how feelings could not only drive the first flush of cultures but remain integral to their evolution, I searched for a way to connect human life, as we know it today—equipped with minds, feelings, consciousness, memory, language, complex sociality, and creative intelligence—with early life, as early as 3.8 billion years ago. To establish the connection, I needed to suggest an order and a time line for the development and appearance of these critical faculties in the long history of evolution.

The actual order of appearance of biological structures and faculties that I uncovered violates traditional expectations and is as strange as the book title implies. In the history of life, events did not comply with the conventional notions that we humans have formed for how to build the beautiful instrument I like to call a cultural mind.

Intending to tell a story about the substance and consequences of human feeling, I came to recognize that our ways of thinking about minds and cultures are out of tune with biological reality. When a living organism behaves intelligently and winningly in a social setting, we assume that the behavior results from foresight, deliberation, complexity, all with the help of a nervous system. It is now clear, however, that such behaviors could also have sprung from the bare and spare equipment of a single cell, namely, in a bacterium, at the dawn of the biosphere. "Strange" is too mild a word to describe this reality.

We can envision an explanation that begins to accommodate the counterintuitive findings. The explanation draws on the mechanisms of life itself and on the conditions of its regulation, a collection of phenomena that is generally designated by a single word: *homeostasis.* Feelings are the mental expressions of homeostasis, while homeostasis, acting under the cover of feeling, is the functional thread that links early life-forms to the extraordinary partnership of bodies and nervous systems. That partnership is responsible for the emergence of conscious, feeling minds that are, in turn, responsible for what is most distinctive about humanity: cultures and civilizations. Feelings are at the center of the book, but they draw their powers from homeostasis.

Connecting cultures to feeling and homeostasis strengthens their links to nature and deepens the humanization of the cultural process. Feelings and creative cultural minds were assembled by a long process in which genetic selection guided by homeostasis played a prominent role. Connecting cultures to feelings, homeostasis, and genetics counters the growing detachment of cultural ideas, practices, and objects from the process of life.

It should be evident that the connections I am establishing do not diminish the autonomy that cultural phenomena acquire historically. I am not reducing cultural phenomena to their biological roots or attempting to have science explain all aspects of the cultural process. The sciences alone cannot illuminate the entirety of human experience without the light that comes from the arts and humanities.

Discussions about the making of cultures often agonize over two conflicting accounts: one in which human behavior results from autonomous cultural phenomena, and another in which human behavior is the consequence of natural selection as conveyed by genes. But there is no need to favor one account over the other. Human behavior largely results from *both* influences in varying proportions and order.

Curiously, discovering the roots of human cultures in nonhuman biology does not diminish the exceptional status of humans at all. The exceptional status of each human being derives from the unique significance of suffering and flourishing in the context of our remembrances of the past and of the memories we have constructed of the future we incessantly anticipate.

3

We humans are born storytellers, and we find it very satisfying to tell stories about how things began. We have reasonable success when the thing to be storied is a device or a relationship, love affairs and friendships being great themes for stories of origins. We are not so good and we are often wrong when we turn to the natural world. How did life begin? How did minds, feelings, or consciousness begin? When did social behaviors and cultures first appear? There is nothing easy about such an endeavor. When the laureate physicist Erwin Schrödinger turned his attention to biology and wrote his classic book *What Is Life?*, it should be noted that he did not title it *The "Origins" of Life*. He recognized a fool's errand when he saw it.

Still, the errand is irresistible. This book is dedicated to presenting some facts behind the making of minds that think, create narratives and meaning, remember the past and imagine the future; and to presenting some facts behind the machinery of feeling and consciousness responsible for the reciprocal connections among minds, the outside world, and its respective life. In their need to cope with the human heart in conflict, in their desire to reconcile the contradictions posed by suffering, fear, anger, and the pursuit of well-being, humans turned to wonder and awe and discovered music making, dancing, painting, and literature. They continued their efforts by creating the often beautiful and sometimes frayed epics that go by such names as religious belief, philosophical inquiry, and political governance. From cradle to grave, these were some of the ways in which the cultural mind addressed the human drama.

PART I

ABOUT LIFE AND ITS REGULATION
(HOMEOSTASIS)

ON THE HUMAN CONDITION

A Simple Idea

When we are wounded and suffer pain, no matter the cause of the wound or the profile of the pain, we can do something about it. The range of situations that can cause human suffering includes not only physical wounds but the sorts of hurts that result from losing someone we love or being humiliated. The abundant recall of related memories sustains and amplifies suffering. Memory helps project the situation into the imagined future and lets us envision the consequences.

Humans would have been able to respond to suffering by attempting to understand their plight and by inventing compensations, corrections, or radically effective solutions. Along with suffering pain, humans were able to experience its very opposites, pleasure and enthusiasm, in a wide variety of situations, ranging from the simple and trivial to the sublime, from the pleasures that constitute responses to tastes and smells, food, wine, sex, and physical comforts, to the wonder of play, the awe and flourishing that arise from the contemplation of a landscape or the admiration and deep affection for another person. Humans also discovered that exerting power, dominating and even destroying others, and causing pure mayhem and pillage could produce not only strategically valuable results but also pleasure. Here, too, humans would have been able to use the existence of such feelings for a practical purpose: as a motive for questioning why pain exists in the first place and perhaps to puzzle at the bizarre fact that under certain circumstances the suffering of others could be rewarding. Perhaps they would have used the related feelings—among them fear, surprise, anger, sadness, and compassion—as a guide to imagining ways of countering suffering and its sources. They would have realized that among the variety of social behaviors available to them, some—fellowship, friendship, care, love—were the very opposite of aggression and violence and were transparently associated with the well-being of not only others but their own.

Why would feelings succeed in moving the mind to act in such an advantageous manner? One reason comes from what feelings accomplish *in* the mind and do *to* the mind. In standard circumstances, feelings tell the mind, without any word being spoken, of the good or bad direction of the life process, at any moment, within its respective body. By doing so, feelings naturally qualify the life process as conducive or not to well-being and flourishing.[1]

Another reason why feelings would succeed where plain ideas fail has to do with the unique nature of feelings. Feelings are not an independent fabrication of the brain. They are the result of a cooperative partnership of body and brain, interacting by way of free-ranging chemical molecules and nerve pathways. This particular and overlooked arrangement guarantees that feelings disturb what might otherwise be an indifferent mental flow. The source of feeling is life on the wire, balancing its act between flourishing and death. As a result, feelings are mental stirrings, troubling or glorious, gentle or intense. They can stir us subtly, in an intellectualized sort of way, or intensely and noticeably, grabbing the owner's attention firmly. Even at their most positive, they tend to disturb the peace and break the quiet.[2]

The simple idea, then, is that feelings of pain and feelings of pleasure, from degrees of well-being to malaise and sickness, would have been the catalysts for the processes of questioning, understanding, and problem solving that most profoundly distinguish human minds from the minds of other living species. By questioning, understanding, and problem solving, humans would have been able to develop intriguing solutions for the predicaments of their lives *and* to construct the means to promote their flourishing. They would have perfected ways of nourishing, clothing, and sheltering themselves, nursing their physical wounds, and beginning the invention of what became medicine. When the pain and the suffering were caused by others—by how they felt about others, by how they perceived others to feel about them—or when the pain was caused by considering their own conditions, such as confronting the inevitability of death, humans would have drawn on their expanding individual and collective resources and invented a variety of responses that ranged from moral prescriptions and principles of justice to modes of social organization and governance, artistic manifestations, and religious beliefs.

It is not possible to tell exactly when these developments would have taken place. Their pace varied significantly depending on the specific populations and their geographic location. We know for certain that by 50,000 years ago such processes were well under way around the Mediterranean, in central and southern Europe, and in Asia, regions where *Homo sapiens* was present, though not without the company of Neanderthals. This was long after *Homo sapiens* first appeared, about 200,000 years ago or earlier.[3] Thus we can think of the beginnings of human cultures as occurring among hunter-gatherers, well before the cultural invention known as agriculture, about 12,000 years ago, and before the invention of writing and money. The dates by which writing systems emerged in varied places are a good illustration of how multi-centered were the processes of cultural evolution. Writing was first developed in Sumer (in Mesopotamia) and in Egypt, between 3500 and 3200 B.C. But a different writing system was later developed in Phoenicia and eventually used by Greeks and Romans. About 600 B.C., writing also developed independently in Mesoamerica, under the Mayan civilization, in the region of contemporary Mexico.

We can thank Cicero and ancient Rome for the word "culture" applied to the universe of ideas. Cicero used the term to describe the cultivation of the soul—*cultura animi*—and he must have been thinking of the tilling of the land and its result, the perfecting and improvement of plant growth. What applied to the land might as well apply to the mind.

There is little doubt about the principal meaning of the word "culture" today. Dictionaries tell us that "culture" refers to manifestations of intellectual achievement regarded collectively, and unless otherwise specified, the word refers to *human* culture. The arts, philosophical inquiry, religious beliefs, moral faculties, justice, political governance, economic institutions—markets, banks—technology, and science are the main categories of endeavor and achievement that are conveyed by the word "culture." The ideas, attitudes, customs, manners, practices, and institutions that distinguish one social group from another belong to the overall scope of culture as does the notion that cultures are transmitted across peoples and generations by language and by the very objects and rituals that the cultures created in the first place. Whenever I refer to cultures or to the cultural mind in this book, this is the scope of phenomena I am considering.

There is another common usage of the word "culture." Amusingly, it refers to the laboratory cultivation of microorganisms such as bacteria: it alludes to bacteria *in* culture, not to the culture-like behaviors of bacteria that we will discuss in a moment. One way or another, bacteria were fated to be part of the grand story of culture.

FEELINGS AND THE MAKING OF CULTURES

Feelings contribute in three ways to the cultural process:

1. as *motives* of the intellectual creation

   a) by prompting the detection and diagnosis of homeostatic deficiencies;

   b) by identifying desirable states worthy of creative effort;

2. as *monitors* of the success or failure of cultural instruments and practices;

3. as participants in the *negotiation* of adjustments required by the cultural process over time.

Feeling Versus Intellect

Conventionally, the human cultural enterprise is explained in terms of exceptional human intellect, a brilliant extra feather in the cap of organisms assembled by unthinking genetic programs over evolutionary time. Feelings rarely earn a mention. The expansion of human intelligence and language, and the exceptional degree of human sociality, are the stars of cultural development. At first glance, there are good reasons to accept this account as reasonable. It is unthinkable to explain human cultures without factoring in the intelligence behind the novel instruments and practices we call culture. It goes without saying that the contributions of language are decisive for the development and transmission of cultures. As for sociality, a contributor that was often ignored, its indispensable role is now apparent. Cultural practices depend on social phenomena at which human adults excel—for example, how two individuals jointly contemplating the same object share an intention regarding that object.[4] And yet something seems to be missing from the intellectual account. It is as if creative intelligence would have materialized without a powerful prompt and would have marched along without a background motive besides pure reason. Presenting survival as a motive will not do because it removes the reasons why survival would be a matter of concern. It is as if creativity would not be embedded in the complex edifice of affect. It is also as if the continuation and monitoring of the process of cultural invention would have been possible by cognitive means alone, without the actual *felt* value of life outcomes, good or bad, having a say in the proceedings. If your pain is medicated with treatment A or treatment B, you rely on feelings to declare which treatment makes the pain less intense, or fully resolved, or unchanged. Feelings work as *motives* to respond to a problem and as *monitors* of the success of the response or lack thereof.

Feelings, and more generally affect of any sort and strength, are the unrecognized presences at the cultural conference table. Everyone in the room senses their presence, but with few exceptions no one talks to them. They are not addressed by name.

In the complementary picture that I am drawing here, exceptional human intellect, individually and socially, would not have been moved to invent intelligent cultural practices and instruments without powerful justifications. Feelings of every sort and shade, caused by actual or imagined events, would have provided the motives and recruited the intellect. Cultural responses would have been created by human beings intent on changing their life situation for the better, for the more comfortable, for the more pleasant, for the more conducive to a future with well-being and with fewer of the troubles and losses that would have inspired such creations in the first place, ultimately and practically, not just for a more survivable future but for a better lived one.

The humans who first devised the Golden Rule, that we should treat others the way we want others to treat us, formulated the precept with the help of what they felt when they were treated badly or when they saw others badly treated. Logic played a role as it worked on facts, to be sure, but some of the critical facts were feelings.

Suffering or flourishing, at the polar ends of the spectrum, would have been prime motivators of the creative intelligence that produced cultures. But so would the experiences of affects related to fundamental desires—hunger, lust, social fellowship—or to fear, anger, the desire for power and prestige, hatred, the drive to destroy opponents and whatever they owned or collected. In fact, we find affect behind many aspects of sociality, guiding the constitution of groups small and large and manifesting itself in the bonds that individuals created around their desires and around the wonder of play, as well as behind conflicts over resources and mates, which were expressed in aggression and violence.

Other powerful motivators included the experiences of elevation, awe, and transcendence that arise from the contemplation of beauty, natural or crafted, from the prospect of finding the means to make ourselves and others prosper, from arriving at a possible solution of metaphysical and scientific mysteries, or, for that matter, from the sheer confrontation with mysteries unsolved.

How Original Was the Human Cultural Mind?

Several intriguing questions arise at this point. On the face of what I have just written, the cultural enterprise originated as a human project. But are the problems that cultures solve exclusively human, or do they concern other living beings as well? And what about the solutions that the human cultural mind advances? Are they a completely original human invention, or were they used, at least in part, by beings that preceded us in evolution? The confrontation with pain, suffering, and the certainty of death, contrasted with the unattained possibility of well-being and flourishing, could well have been—most certainly was—behind some of the creative human processes that gave rise to the now staggeringly complex instruments of culture. But is it not the case that such human constructions were assisted by older biological strategies and instruments that preceded them? When we observe the great apes, we sense the presence of precursors to our cultural humanity. It is known that Darwin was astonished when, in 1838, he first observed the behaviors of Jenny, an orangutan that had recently arrived in the London Zoo. So was Queen Victoria. She found Jenny to be "disagreeably human."[5] Chimpanzees can create simple tools, use them intelligently to feed themselves, and even visually transmit the invention to others. Some aspects of their social behaviors (and those of bonobos in particular) are arguably cultural. So are behaviors of species as far apart as elephants and marine mammals. Thanks to genetic transmission, mammals possess an elaborate affective apparatus that, in many respects, resembles ours in its emotional roster. To deny mammals the feelings related to their emotionality is no longer a tenable position. Feelings could also have played a motivating role to account for the "cultural" manifestations of nonhumans. Importantly, the reason why their cultural achievements turned out to be so modest would be related to the lesser development or absence of traits such as shared intentionality and verbal language, and, more generally, the modesty of their intellect.

But things are not so simple. Given the complexity and wide-ranging positive and negative consequences of cultural practices and tools, it would be reasonable to expect that their conception would have been intentional and possible only in minded creatures, as non-human primates certainly are, perhaps after a holy alliance of feeling and creative intelligence could devote itself to the problems raised by existence in a group. Before cultural manifestations could emerge in evolution, one would first have had to wait for the evolutionary development of minds and feeling—complete with consciousness, so that feeling could be experienced subjectively—and then wait some more for the development of a healthy dose of mind-directed creativity. So goes the conventional wisdom, but that is not true as we are about to see.

Humble Beginnings

Social governance has humble beginnings, and neither the minds of *Homo sapiens* nor of other mammalian species were present at its natural birth. Very simple unicellular organisms relied on chemical molecules to *sense and respond,* in other words, to detect certain conditions in their environments, including the presence of others, and to guide the actions that were needed to organize and maintain their lives in a social environment. It is known that bacteria growing in fertile terrain, rich in the nutrients they need, can afford to live relatively independent lives; bacteria living in terrain where nutrients are scarce band together in clumps. Bacteria can sense the numbers in the groups they form and in an unthinking way assess group strength, and they can, depending on the strength of the group, engage or not in a battle for the defense of their territory. They can physically align themselves to form a palisade, and they can secrete molecules that constitute a thin veil, a film that protects their ensemble and probably plays a role in the bacteria's resistance against the action of antibiotics. By the way, this is what goes on routinely in our throats when we get a cold and develop pharyngitis or laryngitis. When bacteria gain a lot of throat territory, we become hoarse and lose our voices. "Quorum sensing" is the process that assists bacteria in these adventures. The achievement is so spectacular that it makes one think of capabilities such as feeling, consciousness, and reasoned deliberation, except that bacteria do not have any such capabilities; they have rather the powerful *antecedents* to those capabilities. I will argue that they lack the mental expression of those antecedents. Bacteria do not engage in phenomenology.[6]

Bacteria are the earliest form of life, dating back to almost four billion years ago. Their body consists of one cell, and the cell does not even have a nucleus. They have no brain. They have no mind in the sense that I and the reader do. They appear to lead a simple life, operating according to the rules of homeostasis, but there is nothing simple about the flexible chemistries that they operate and that allow them to breathe the unbreathable and eat the uneatable.

In the complex, albeit un-minded, social dynamic they create, bacteria can cooperate with other bacteria, genomically related or not. And in their un-minded existence, it turns out they even assume what can only be called a sort of "moral attitude." The closest members of their social group, their family so to speak, are mutually identifiable by the surface molecules they produce or chemicals they secrete, which are in turn related to their individual genomes. But groups of bacteria have to cope with the adversity of their environments and often have to compete with other groups in order to gain territory and resources. For a group to be successful, its members need to cooperate. What can happen during the group effort is fascinating. When bacteria detect "defectors" in their group, which really means that certain members fail to help with the defense effort, they shun them even if they are genomically related and therefore part of their family. Bacteria will not cooperate with kin bacteria that do not pull their weight and help with the group endeavor; in other words, they snub noncooperative turncoat bacteria. Cheaters, after all, gain access to energy resources and defense that the rest of the group is providing at great cost, at least for a while. The variety of possible bacterial "conduct" is remarkable.[7] In one telling experiment designed by the microbiologist Steven Finkel, several populations of bacteria were to fend for resources inside flasks equipped with different proportions of the necessary nutrients. In one particular condition, over multiple generations, the experiment revealed three distinct successful bacteria groups: two that fought each other to the death and suffered major losses in the process, and one that had sailed discreetly over time, without any frontal engagement. All three groups made it into the future, a future as long as twelve thousand generations. We do not have to be very imaginative to sense comparable patterns in big-creature societies. Societies of cheaters or of peaceable, law-abiding citizens come to mind. It is easy to conjure up a colorful cast of characters, abusers, bullies, thugs, and thieves, but also quiet dissimulators who do very well, just not brilliantly, and, last but not least, the wonderful altruists.[8]

One would be very foolish to reduce the sophistication of humanly developed moral rules and application of justice to the spontaneous behavior of bacteria. We should not confuse the formulation and thoughtful application of a rule of law with the strategy schema used by bacteria when they end up joining forces with a cooperative non-kin, the usual enemy, instead of the kin, their usual friend. In their un-minded orientation to survival, they join with others working toward the same goal. Following the same undeliberated rule, the group response to overall attacks consists of automatically seeking strength in numbers following the equivalent of the principle of least action.[9] Their obeisance of homeostatic imperatives is strict. Moral principles and the law obey the same core rules, but not only. Moral principles and laws are the result of intellectual analyses of the conditions humans have faced and of the management of power by the group inventing and promulgating laws. They are grounded in feeling, knowledge, and reasoning, processed in a mental space, with the use of language.

One would be equally foolish, however, not to recognize that simple bacteria have governed their lives for billions of years according to an automatic schema that foreshadows several behaviors and ideas that humans have used in the construction of cultures. Nothing in our human conscious minds tells us overtly that these strategies have existed for so long in evolution or when they first appeared, although when we introspect and search our minds for how we should act, we do find "hunches and tendencies," hunches and tendencies that are informed by feelings or *are* feelings. Those feelings gently or forcefully guide our thoughts and actions in a certain direction, providing scaffolding for intellectual elaborations and even suggesting justifications for our actions: for example, welcoming and embracing those who help us when we are in need; shunning those who are indifferent to our plight; punishing those who abandon us or betray us. But we would never have known that bacteria do smart things that work in the same direction without the current science that has so revealed. Our natural behavioral tendencies have guided us toward a conscious elaboration of basic and nonconscious principles of cooperation and struggle that have been present in the behavior of numerous forms of life. Those principles have also guided, over long spans of time and in numerous species, the evolutionary assembly of affect and its key components: all the emotive responses generated by sensing varied internal and external stimuli that engage appetitive drives—thirst, hunger, lust, attachment, care, fellowship—and recognizing situations that require emotional responses such as joy, fear, anger, and compassion. Those principles, which as noted earlier are easily recognizable in mammals, are ubiquitous in the history of life. It is apparent that natural selection and genetic transmission have been hard at work in shaping and sculpting such modes of reacting in social environments to construct the scaffolding of the human cultural mind. Together, subjective feelings and creative intelligence have operated in that setting and created cultural instruments that serve the needs of our lives. If that is indeed the case, the human unconscious literally goes back to early life-forms, deeper and further than Freud or Jung ever dreamed.

From the Life of Social Insects

Now consider this. A small number of invertebrate species, a mere 2 percent of all species of insects, is capable of social behaviors that do rival in complexity many human social achievements. Ants, bees, wasps, and termites are the prominent examples.[10] Their genetically set and inflexible routines enable the survival of the group. They divide labor intelligently within the group to deal with the problems of finding energy sources, transform them into products useful for their lives, and manage the flow of those products. They do so to the point of changing the number of workers assigned to specific jobs depending on the energy sources available. They act in a seemingly altruistic manner whenever sacrifice is needed. In their colonies, they build nests that constitute remarkable urban architectural projects and provide efficient shelter, traffic patterns, and even systems of ventilation and waste removal, not to mention a security guard for the queen. One almost expects them to have harnessed fire and invented the wheel. Their zeal and discipline put to shame, any day, the governments of our leading democracies. These creatures acquired their complex social behaviors from their biology, not from Montessori schools or Ivy League colleges. But in spite of having come by these astounding abilities as early as 100 million years ago, ants and bees, individually or as colonies, do not grieve for the loss of their mates when they disappear and do not ask themselves about their place in the universe. They do not inquire about their origin, let alone their destiny. Their seemingly responsible, socially successful behavior is not guided by a sense of responsibility, to themselves or to others, or by a corpus of philosophical reflections on the condition of being an insect. It is guided by the gravitational pull of their life regulation needs as it acts on their nervous systems and produces certain repertoires of behavior selected over numerous evolving generations, under the control of their fine-tuned genomes. Members of a colony do not think as much as they act, by which I mean that upon registering a particular need—theirs, or the group's, or the queen's—they do not ponder alternatives for how to fulfill such a need in any way comparable to ours. They simply fulfill it. Their repertoire of actions is limited, and in many instances it is confined to one option. The general schema of their elaborate sociality does resemble that of human cultures, but it is a fixed schema. E. O. Wilson calls social insects "robotic" and for good reason.

Now, back to humans. We humans do ponder alternatives for our behavior, do mourn the loss of others, do want to do something about our losses and about maximizing our gains, and do ask questions about our origin and destiny and propose answers, and we are so disorderly in our bubbling and conflicting creativities that we are often a mess. We do not know exactly when humans began grieving, reacting to losses and gains, commenting on their condition, and asking inconvenient questions about the wherefrom and whereto of their lives. We know for certain, based on artifacts from the burial sites and caves that have been explored to date, that 50,000 years ago some of these processes were well established. But note how, amazingly, this is a mere evolutionary instant when we compare, say, 50 *thousand* years of humanity to 100 *million* years of the lives of social insects, not to mention a few *billion* years of history for bacteria.

Although we do not descend directly from bacteria or social insects, I believe it is instructive to reflect on these three lines of evidence: bacteria devoid of brains or minds that defend their turf, wage warfare, and act according to something equivalent to a code of conduct; enterprising insects that create cities, systems of governance, and functional economies; and humans who invent flutes, write poetry, believe in God, conquer the planet and the space around, fight diseases to alleviate suffering but also destroy other humans for their own gain, invent the Internet, find ways to turn it into an instrument of progress and of catastrophe, and, to boot, ask questions about bacteria, about ants and bees, and about themselves.

Homeostasis

How can we reconcile the seemingly reasonable idea that feelings motivated intelligent cultural solutions for problems posed by the human condition with the fact that un-minded bacteria exhibit socially effi-cacious behaviors whose contours foreshadow some human cultural responses? What is the thread that links these two sets of biological manifestations, whose emergence is separated by billions of years of evolution? I believe that the common ground and the thread can be found in the dynamics of *homeostasis*.

Homeostasis refers to the fundamental set of operations at the core of life, from the earliest and long-vanished point of its beginning in early biochemistry to the present. Homeostasis is the powerful, unthought, unspoken imperative, whose discharge implies, for every living organism, small or large, nothing less than enduring and pre-vailing. The part of the homeostatic imperative that concerns "endur-ing" is transparent: it produces survival and is taken for granted without any specific reference or reverence whenever the evolution of any organism or species is considered. The part of homeostasis that concerns "prevailing" is more subtle and rarely acknowledged. It ensures that *life is regulated within a range that is not just compatible with survival but also conducive to flourishing, to a projection of life into the future of an organism or a species.*

Feelings are the very revelation to each individual mind of the sta-tus of life within the respective organism, a status expressed along a range that runs from positive to negative. Deficient homeostasis is expressed by largely negative feelings, while positive feelings express appropriate levels of homeostasis and open organisms to advanta-geous opportunities. Feelings and homeostasis relate to each other closely and consistently. Feelings are the subjective experiences of the state of life—that is, of homeostasis—in all creatures endowed with a mind and a conscious point of view. We can think of feelings as mental deputies of homeostasis.[11]

I bemoaned the neglect of feelings in the natural history of cul-tures, but the situation is even worse in relation to homeostasis and life itself. Homeostasis and life are left out altogether. Talcott Parsons, one of the most prominent sociologists of the twentieth century, did invoke the notion of homeostasis in relation to social systems, but in his hands the concept was not connected to life or feeling. Parsons is actually a good example of the neglect of feeling in the conception of cultures. For Parsons, the brain was the organic foundation of culture because it was the "primary organ for controlling complex operations, notably manual skills, and coordinating visual and auditory information." Above all, the brain was "the organic basis of the capacity to learn and manipulate symbols."[12]

Homeostasis has guided, non-consciously and non-deliberatively, without prior design, the selection of biological structures and mechanisms capable of not only maintaining life but also advancing the evolution of species to be found in varied branches of the evolutionary tree. This conception of homeostasis, which conforms most closely to the physical, chemical, and biological evidence, is remarkably different from the conventional and impoverished conception of homeostasis that confines itself to the "balanced" regulation of life's operations.

It is my view that the unshakable imperative of homeostasis has been the pervasive governor of life in all its guises. Homeostasis has been the basis for the value behind natural selection, which in turn favors the genes—and consequently the kinds of organisms—that exhibit the most innovative and efficient homeostasis. The development of the genetic apparatus, which helps regulate life optimally and transmit it to descendants, is not conceivable without homeostasis.

Given the foregoing, we can advance a working hypothesis on the relation between feelings and cultures. *Feelings, as deputies of homeostasis, are the catalysts for the responses that began human cultures.* Is this reasonable? Is it conceivable that feelings could have motivated the intellectual inventions that gave humans (1) the arts, (2) philosophical inquiry, (3) religious beliefs, (4) moral rules, (5) justice, (6) political governance systems and economic institutions, (7) technology, and (8) science? I would respond yes, wholeheartedly. I can make the case that cultural practices or instruments in each of the eight areas above required feeling a situation of actual or anticipated homeostatic decline (for example, pain, suffering, dire need, threat, loss) or of potential homeostatic benefit (for example, a rewarding outcome), and that feeling acted as the motive for exploring, using the instruments of knowledge and reason, the possibilities of reducing a need or of capitalizing on the abundance signified by reward states.

But this is only the beginning of the story. The consequence of a successful cultural response is the decline or canceling of the motivating feeling, a process that requires *monitoring* changes, in homeostatic status. In turn, the eventual adoption of the actual intellectual responses and their inclusion in a cultural corpus—or their abandonment—are a complex process resulting from interactions of varied social groups over time. It depends on numerous characteristics of the groups, from size and past history to geographic location and internal and external power relationships. It involves subsequent intellectual and feeling steps, for example, when cultural conflicts emerge, negative as well as positive feelings are engaged and contribute to either solving or aggravating the conflicts. It makes use of cultural selection.

Foreshadowing Minds and Feelings Is Not the Same as Generating Minds and Feelings

Life would not be viable without the traits imposed by homeostasis, and we know that homeostasis has existed ever since life began. But feelings—the subjective experiences of the momentary state of homeostasis within a living body—did not emerge when life did. I propose that they emerged only after organisms were endowed with nervous systems, a far more recent development that began to occur only about 600 million years ago.

Nervous systems gradually enabled a process of multidimensional mapping of the world around them, a world that begins in the organism's interior, so that minds—and feelings within those minds—would be possible. The mapping was based on varied sensory abilities, which eventually came to include smell, taste, touch, hearing, and vision. As will become clear in chapters 4 through 9, the making of minds—and of feelings in particular—is grounded on *interactions* of the nervous system and its organism. *Nervous systems make minds not by themselves but in cooperation with the rest of their own organisms.* This is a departure from the traditional view of brains as the sole source of minds.

Although the emergence of feelings is far more recent than the beginnings of homeostasis, it still occurred long before humans entered the scene. Not all creatures are endowed with feelings, but *all* living creatures are equipped with the regulation devices that were precursors to feelings (some of which are discussed in chapters 7 and 8).

As we consider the behavior of bacteria and social insects, suddenly early life is modest in name only. The actual beginnings of what eventually became human life, human cognition, and the cast of mind that I like to call cultural go back to a vanishing point in the history of the earth. It is not enough to say that our minds and cultural successes are grounded in brains that share numerous features with the brains of our mammalian relatives. We have to add that our minds and cultures are linked to the ways and means of ancient unicellular life and of many intermediate life-forms. One might say, figuratively, that our minds and cultures have borrowed from the past liberally, without embarrassment or apology.

Early Organisms and Human Cultures

It is important to insist that identifying links between biological processes, on the one hand, and mental and sociocultural phenomena, on the other, does not signify that the shape of societies and the makeup of cultures can be fully explained by the biological mechanisms we are outlining. Certainly I suspect that the development of codes of conduct, regardless of where or when they appeared, has been inspired by the homeostatic imperative. Such codes have generally aimed at the reduction of risks and dangers for individuals and social groups and have indeed resulted in a reduction of suffering and the promotion of human welfare. They have strengthened social cohesion, which is, in and of itself, favorable to homeostasis. But beyond the fact that they were conceived by humans, Hammurabi's Code, the Ten Commandments, the U.S. Constitution, and the Charter of the United Nations were shaped by specifics of the circumstances of their time and place and by the particular humans who developed such codes. There are several formulas behind such developments rather than one single comprehensive formula, although parts of any of the possible formulas are universal.

Biological phenomena can prompt and shape events that become cultural phenomena, and must have done so at the dawn of cultures via the interplay of affect and reason, in specific circumstances defined by the individuals, the groups, their location, their past, and so forth. Also, the intervention of affect was not confined to an initial motive. It recurred in the role of monitor of the process and continued to intervene into the future of many cultural inventions as required by the everlasting negotiations between affect and reason. But the critical biological phenomena—feelings and intellect within cultural minds— are only one part of the story. Cultural selection needs to be factored in, and to do so we require the scholarship of history, geography, and sociology, among many other disciplines. At the same time, we need to recognize that the adaptations and faculties used by cultural minds were the result of natural selection and genetic transmission.

Genes were instrumental in the traversals from early life to human life today. That much is obvious and true but begs the question of how genes came to be and to do so. A more complete answer, perhaps, is that even at the earliest, long-vanished point, the physical and chemical conditions of the life process were responsible for establishing homeostasis in the ample sense of the term and everything else flowed from that fact, including the machinery of genes. This happened in cells without nuclei (or prokaryotes). Later, homeostasis was behind the selection of cells with nuclei (or eukaryotes). Later still come complex organisms with many cells. Eventually, such multicellular organisms elaborated existing "whole-body systems" into the endocrine, immune, circulatory, and nervous systems. Such systems gave rise to minds, feelings, consciousness, the machinery of affect, and complex movements. Without such whole-body systems, multicellular organisms would not have been able to operate their "global" homeostasis.

The brains that have helped human organisms invent cultural ideas, practices, and instruments were assembled by genetic inheritance, naturally selected over billions of years. By contrast, the products of the human cultural mind and the history of humans have been subject mostly to cultural selection and have been transmitted to us largely by cultural means.

In the march toward the human cultural mind, the presence of feelings would have allowed homeostasis to make a dramatic leap because they could represent mentally the state of life within the organism. Once feelings were added to the mental mix, the homeostatic process was enriched by direct knowledge of the state of life and, of necessity, that knowledge was conscious. Eventually, each feeling-driven, conscious mind could mentally represent, with an explicit reference to the experiencer subject, two critical sets of facts and events: (1) the conditions in the inner world of its own organism; and (2) the conditions of its organism's environment. The latter prominently included the behaviors of other organisms in a variety of complex situations generated by social interactions as well as by shared intentions, many of them dependent on the individual drives, motivations, and emotions of the participants.

As learning and memory advanced, individuals became able to establish, recall, and manipulate memories of facts and events, opening the way to a new level of intelligence based on knowledge and feeling. Into this process of intellectual expansion came verbal language, providing easily manipulable and transmissible correspondences between ideas and words and sentences. From there on, the creative flood could not be contained. Natural selection had just conquered yet another theater of operations, that of the ideas behind certain actions, practices, and artifacts. Cultural evolution could now join genetic evolution.

The prodigious human mind and the complicated brain that enables it distract us from the long line of biological antecedents that account for their presence. The splendor of mind and brain achievements makes it possible to imagine that human organisms and minds could spring forth fully formed, like a phoenix, parentage unknown or very recent. Behind such prodigies, however, there are long chains of precedents and amazing degrees of competition and cooperation. How easy it is to overlook, in the story of our minds, the fact that life in complex organisms could only have endured and prevailed if it were curated and that brains came to be favored in evolution because they became so good at assisting with the curatorial job, especially after they were able to help organisms fabricate conscious minds rich in feeling and thinking. In the end, human creativity is rooted in life and in the breathtaking fact that life comes equipped with a precise mandate: resist and project itself into the future, no matter what. It may be helpful to consider these humble but powerful origins as we cope with the instabilities and uncertainties of the present.

Contained within life's imperative and its seeming homeostatic magic, coiled as it were, there were instructions for immediate survival: the regulation of metabolism and repair of cellular components, rules for behavior in a group, and standards for the measurement of positive and negative departures from homeostatic balance so that appropriate responses could be launched. But the imperative also harbored the tendency to seek future security in more complex and robust structures, a relentless plunge into the future. The realization of this tendency was achieved by myriad cooperations, along with the mutations, and fierce competition that enabled natural selection. Early life was foreshadowing many future developments that we can now observe in human minds imbued with feeling and consciousness and enriched by the cultures that such minds have constructed. Complex, conscious, feeling minds inspired and steered the expansion of intelligence and language and generated novel instruments of dynamic homeostatic regulation external to living organisms. The intentions expressed by such new instruments are still consonant with the early life imperative, still aimed at not just enduring but prevailing.

Why, then, are the results of these extraordinary developments so inconsistent, not to say erratic? Why so much derailed homeostasis and so much suffering over human history? A preliminary answer, which we will address later in the book, is that cultural instruments first developed in relation to the homeostatic needs of individuals and of groups as small as nuclear families and tribes. The extension to wider human circles was not and could not have been contemplated. Within wider human circles, cultural groups, countries, even geopolitical blocs, often operate as individual organisms, not as parts of one larger organism, subject to a single homeostatic control. Each uses the respective homeostatic controls to defend the interests of *its* organism. Cultural homeostasis is merely a work in progress often undermined by periods of adversity. We might venture that the ultimate success of cultural homeostasis depends on a fragile civilizational effort aimed at reconciling different regulation goals. This is why the calm desperation of F. Scott Fitzgerald—"so we beat on, boats against the current, borne back ceaselessly into the past"—remains a prescient and appropriate way of describing the human condition.[13]

IN A REGION OF UNLIKENESS

Life

Life, at least the life that we descend from, appears to have begun about 3.8 billion years ago, long after the Big Bang of great fame, quietly, with discretion, no fanfare to herald its now astonishing onset, on planet Earth, under the protection of our own sun, in the general department of the Milky Way.

Present were the crust of Earth, its oceans and atmosphere, particular conditions of the environment such as temperature, and certain critical elements—carbon, hydrogen, nitrogen, oxygen, phosphorus, and sulfur.

Protected by a circumventing membrane, a number of processes emerged within a set-aside region of unlikeness known as a cell.[1] Life began within that first cell—*was* that cell—as an extraordinary assembly of chemical molecules with particular affinities and ensuing self-perpetuating chemical reactions, ticking, beating, repeating cycles. On its own and of its own accord, the cell repaired the inevitable wear and tear. When one part broke, the cell substituted it, more or less exactly, and so the cell's functional arrangements were maintained and life continued unabated. "Metabolism" is the single name for the chemical pathways that accomplished this feat, a process that required the cell to extract, as efficiently as possible, the necessary energy from sources in its environment, use the energy, equally efficiently, for the purpose of rebuilding the broken machinery, and throw away the waste products. "Metabolism" is a word of recent coinage (late nineteenth century) derived from the Greek for "change." Metabolism covers the processes of catabolism—a breakdown of molecules that results in the release of energy—and anabolism, a process of construction that consumes energy. The term "metabolism," which is used in English and in the Romance languages, is rather opaque, unlike the equivalent German term—*Stoffwechsel*, or "exchange of stuff, of material." As Freeman Dyson spiritedly points out, the German word suggests what metabolism is all about.[2]

But there was more to the process of life than just an evenhanded maintenance of balance. From a number of possible "steady states," the cell, at the peak of its powers, naturally tended to the steady state most conducive to positive energy balances, a surplus with which life could be optimized and projected into the future. As a result, the cell could flourish. In this context, flourishing signifies both a more efficient way of living and the possibility of reproduction.

The collection of coordinated processes required to execute life's unthought and unwilled desire to persist and advance into the future, through thick and thin, is known as homeostasis. I know that "unwilled," "unthought," and "desired" are seemingly conflicting terms, but in spite of the apparent paradox these are the most convenient ways of describing the process. No exactly comparable process appears to have existed prior to the onset of life, although one can let imagination glean some precursors in the behavior of molecules and atoms. Still, the emergent state of life seems tied to particular kinds of substrate and chemical process. It is reasonable to say that homeostasis has its origins at the cellular, most simple level of life, of which bacteria are prime examples in all their shapes and sizes. Homeostasis refers to the process by which the tendency of matter to drift into disorder is countered so as to maintain order but at a new level, the one allowed by the most efficient steady state. This countering takes advantage of the principle of least action—enunciated by the French mathematician Pierre Maupertuis—whereby free energy will be consumed most efficiently and as fast as possible. Envision the uncanny job of a jongleur, who is not allowed to rest from his endeavor of keeping all balls up in the air and letting none fall, and you get a theatrical representation of the vulnerability and risk of life. And now think that the jongleur also wants to impress you with his elegance and speed, his brilliance, and then you realize that he is already considering an even better act.[3]

In brief, each cell manifested, and all cells forever so, a powerful, seemingly indomitable "intention" to maintain itself alive and to sail forth. That indomitable intention fails only in circumstances of disease or aging, when the cell literally implodes in a process known as apoptosis. Let me stress that I do not think cells have intentions, desires, or wills, in the same way that minded and conscious beings do, but they can behave as if they do and they did. When the reader or I have an intention or desire or will, we can represent several aspects of the process explicitly in *mental* form; individual cells cannot, at least not in the same way. Still, non-consciously, their actions aim at persistence into the future and these actions are the consequences of particular chemical substrates and interactions.

This indomitable intention corresponds to the "force" that the philosopher Spinoza intuited and named the *conatus*. We now understand that it is present at the microscopic scale of each living cell, and we can envision it projected, at the macroscopic scale, everywhere we look in nature: to our whole organisms, made up of trillions of cells, to the billions of neurons in our brains, to the minds that arise in our embodied brains, and to the countless cultural phenomena that the collectives of human organisms have been constructing and tinkering with for millennia.

The continuous attempt at achieving a state of positively regulated life is a defining part of our existence—the first reality of our existence, as Spinoza would say when he described the relentless endeavor of each being to preserve itself. A blend of striving, endeavor, and tendency comes close to rendering the Latin *conatus*, as used by Spinoza in propositions 6, 7, and 8 of the *Ethics*, part 3. In Spinoza's own words, "Each thing, as far as it can be its own power, strives to persevere in its being," and "The striving by which each thing strives to persevere in its being is nothing but the actual essence of the thing." Interpreted with the advantage of current hindsight, Spinoza says that the living organism is constructed so as to maintain the coherence of its structures and functions, for as long as possible, against the odds that threaten it. It is interesting to note that Spinoza reached these conclusions before Maupertuis advanced the principle of least action (Spinoza died almost half a century before). He would have welcomed the support.[4]

In spite of the transformations that the body undergoes as it develops, renews its constituent parts, and ages, the *conatus* insists on maintaining the same individual, respecting the original architectural plan, and thus allowing for the sort of animation that is associated with that plan. The animation can vary in scope, corresponding to life processes merely sufficient to survive or to achieve optimal life processes.

The poet Paul Éluard wrote about the *dur désir de durer*, another way of describing the *conatus* but with the alliterative beauty of a memorable collection of French sounds. I can translate it, pallidly, as the "determined desire to endure." And William Faulkner wrote of the human desire to "endure and prevail." He, too, was referring, with remarkable intuition, to the projection of the *conatus* in the human mind.[5]

Life on the Move

There are plenty of bacteria around us, on us, and inside us, today, but there are no examples left around of those very early bacteria of 3.8 billion years ago. What they were like, what early life was like exactly, needs to be pieced together from different strands of evidence. Between the beginnings and now, there are sparsely documented gaps. How life arose, precisely, is open to informed conjecture.

At first blush, in the wake of the discovery of the structure of DNA, the elucidation of the role of RNA, and the breaking of the genetic code, it must have appeared that life had to come from the genetic material, but that idea was up against a major difficulty: the likelihood of such complex molecules assembling themselves spontaneously as the first step in the construction of life was low to nil.[6]

The puzzlement and equivocation were perfectly understandable. The 1953 discovery (by Francis Crick and James Watson and Rosalind Franklin) of the double-helix structure of DNA was and remains one of the peak moments of the history of science and deservedly influenced the formulations of life that followed. DNA was inevitably seen as the molecule of life and, by extension, the molecule of its beginning. But how could a molecule so complex put itself together spontaneously in the primordial soup? Seen from that perspective, the likelihood of life's spontaneous emergence was so negligible that it justified Francis Crick's skepticism that it would have originated on Earth. He and his colleague Leslie Orgel, at the Salk Institute, thought that life might have come from outer space, brought in by rocket ships, unmanned. This was a version of Enrico Fermi's idea that aliens from other planets would have come to Earth and brought life with them. As intriguing as this claim is, it simply pushed the problem out to another planet. The aliens would have vanished, in the meantime, or perhaps be in our midst but unrecognized. The Hungarian physicist Leo Szilard ventured that of course they were still among us "but they call themselves Hungarians."[7] This is especially amusing because another notable Hungarian, the biologist and chemical engineer Tibor Gánti, was a critic of the idea that life had been shipped from elsewhere, a notion that Crick eventually abandoned.[8] Still the puzzlement over the emergence of life produced widely divergent views, from some of the most distinguished biologists of the twentieth century. Jacques Monod, for example, was a "life skeptic" and believed that the universe was "not pregnant with life," while Christian de Duve thought exactly the opposite.

Today we are still faced with two competing views: one we may call "replicator first" and the other "metabolism first." The replicator-first view is attractive because the machinery of genetics is reasonably well understood and so compelling. When people pause to consider the origin of life, which surprisingly people rarely do, replicator first is the default account. Because genes help manage life and can transmit life, why would they not have started the life ball rolling? Richard Dawkins, for example, favors this view.[9] The primordial soup would beget replicator molecules, which would beget living bodies, which then slave for an assigned lifetime to protect the integrity of genes and their selective, triumphant march along evolution. Stanley Miller and Harold Urey had reported, also in 1953, that the equivalent of a lightning storm inside a test tube could produce amino acids, the building blocks of proteins, thus making simple chemical beginnings plausible.[10] Eventually, elaborate bodies such as ours, equipped with brains and minds and creative intelligence, would come into being to do, once again, the gene's bidding. Whether one finds this account plausible or compelling is a matter of taste. The difficulty is not to be taken lightly, because nothing is that transparent on the issue of life's origins. In favor of this view, a scenario has been advanced in which geological conditions about 3.8 billion years ago would have been compatible with the spontaneous assembly of some of the RNA nucleotides. The RNA world would account for the chemical autocatalytic cycles that define metabolism and for genetic transmission. In a variation on this theme, catalytic RNAs would do double duty, replicate and do the chemistry.

The version of events that I find most persuasive, however, calls for metabolism first. In the beginning, it was plain chemistry, as Tibor Gánti would propose. The primordial soup contained key ingredients, and there were enough favorable conditions, such as thermal vents and lightning storms, you name it, that certain molecules and certain chemical pathways were assembled and initiated their ceaseless protometabolic operations. Living matter would have begun as a chemical sleight of hand, a result of cosmic chemistry and of its inevitability, but living matter would be imbued with the homeostatic imperative, and that would set the agenda. In addition to the forces selecting for increasingly stable molecular and cellular conformations, which achieved life persistence and positive energy balances, there was a set of fortuitous events leading to the generation of self-copying molecules such as nucleic acids. This process achieved two feats: a centrally organized mode of internal life regulation and a mode of genetic transmission of life that superseded simple cell division. The perfecting of the double-tasked genetic machinery would not have stopped since.

This version of life's beginning has been persuasively articulated by Freeman Dyson and is favored by a number of chemists, physicists, and biologists, among them J. B. S. Haldane, Stuart Kauffman, Keith Baverstock, Christian de Duve, and P. L. Luisi. The autonomy of the process, the fact that life is generated from "within," self-started and self-maintained in all of its aspects, was also well captured by the Chilean biologists Humberto Maturana and Francisco Varela in the process that they named *autopoiesis*.[11]

Curiously, on the metabolism-first account, homeostasis "tells" the cell, as it were, to do its business as perfectly as possible so that the *cell's* life can persist. This is the same exhortation that genes are supposed to make to the living cell in the replicator account, except that the goal of genes is their own persistence, not the cell's life. In the end, independent of how exactly things began, the homeostatic imperative manifested itself not only in the metabolic machinery of cells but also in the mechanism of regulation and replication of life. In a world of DNA, two distinct kinds of life—isolated cells and multicellular organisms—were eventually equipped with genetic machinery capable of reproducing themselves and generating offspring, but the genetic apparatus that assisted organisms with reproduction also came to assist them with the fundamental regulation of metabolism.

In a simple way, the region of unlikeness called life, at the level of humble cells—without and with a nucleus—or of large multicellular organisms such as we humans are, can be defined by these two traits: the ability to regulate *its* life by maintaining internal structures and operations for as long as possible, and the possibility of reproducing itself and taking a stab at perpetuity. It is as if, in an extraordinary way, each of us, each cell in us, and every other cell were part of one single, gigantic, supertentacular organism, the one and only organism that began 3.8 billion years ago and still keeps going.

Looking back, this is all in keeping with Erwin Schrödinger's definition of life. Schrödinger, who was a laureate physicist, ventured in 1944 into the realm of biology with remarkable results. His brief masterpiece, titled *What Is Life?*, delivers an anticipation of the likely arrangement of the small molecule required for the genetic code, and his ideas had a major influence on Francis Crick and James Watson. As for the answer to his book's title question, here are some of the key passages.[12]

"Life seems to be orderly and lawful behavior of matter, not based exclusively on its tendency to go over from order to disorder, but based partly on existing order that is kept up." The idea of "existing order that is kept up" is pure Spinoza, the philosopher he quotes at the beginning of his book. The *conatus* is the force that in Schrödinger's words counters "the natural tendency of things to go over into disorder," a resistance that Schrödinger sees expressed in living organisms and in the heredity molecule he was envisioning.

"What is the characteristic feature of life? When is a piece of matter said to be alive?" asks Schrödinger. His answer:

> When it goes on 'doing something,' moving, exchanging material with its environment, and so forth, and that for a much longer period than we would expect in an inanimate piece of matter to 'keep going' under similar circumstances. When a system that is not alive is isolated or placed in a uniform environment, all motion usually comes to a standstill very soon as a result of various kinds of friction; differences of electric or chemical potential are equalized, substances which tend to form a chemical compound do so, temperature becomes uniform by heat conduction. After that the whole system fades away into a dead, inert lump of matter. A permanent state is reached, in which no observable events occur. The physicist calls this the state of thermodynamical equilibrium, or of "maximum entropy."

Well-groomed metabolism—that is, metabolism guided by homeostasis—would define the beginnings of life and its movement forward and be the driving force for evolution. Natural selection, which is guided by the most efficient extraction of nutrients and energy from the environment, did the rest, which included centralized metabolic regulation and replication.

Because nothing like life and its imperative seems to have existed prior to about four billion years ago, when the dissipation of heat produced liquid water, this means that it took almost ten billion years for the right chemistry to appear, in the right spot, not long after Earth was formed and had time to cool off. Then the novelty of life could emerge and begin its relentless course toward complexity and varied species. Whether life exists elsewhere in space remains an open question to be decided by the appropriate exploration. There might even be other kinds of life with a different chemical base. We simply do not know.

We still cannot create life from scratch in a test tube. We know the ingredients of life, we know how genes transmit life to new organisms and how they manage life within the organism, and we are able to create organic chemicals in a laboratory. It is possible to successfully implant a genome in a bacterium from which its own genome has been removed. The newly inserted genome will run the bacterium's homeostasis and allow it to reproduce more or less perfectly. One might say that the new genome is inhabited by its own *conatus* and can deploy its intentions. But creating life from scratch, clean, pre-gene chemical life as it might once have been in that first-ever region of unlikeness, still eludes us.[13]

Organizing chemistry so that it results in life is not for the fainthearted.

Understandably, most conversations about the science of life focus on the amazing machinery of genes, responsible as it currently is to transmit and partly regulate life. But when we talk about life itself, genes are not all there is to talk about. In fact, it is reasonable to hypothesize that the homeostatic imperative, as encountered in the very first life-forms, was followed by the genetic material, not the other way around. This would have been achieved as a result of its constitutive but unspelled endeavor toward optimization of life, which is itself behind natural selection. Genetic material would have assisted the homeostatic imperative to the best advantage: by being responsible for the generation of progeny, which is an attempt at guaranteeing perpetuity, it would have enacted the ultimate consequence of homeostasis.

The biological structures and operations responsible for homeostasis embody the biological value on the basis of which natural selection operates. This wording helps with the issue of origins and locates the critical physiological process in particular conditions of the life process and of its underlying chemistry.

Where genes fit in the history of life is not a trivial issue. Life, its homeostatic imperative, and natural selection point to the appearance of genetic processes and benefit from them. Life, its homeostatic imperative, and natural selection also explain the evolutionary rise of intelligent behaviors, including social behaviors, in single-celled organisms, as well as the eventual rise, in multicellular organisms, of nervous systems and minds imbued with feeling, consciousness, and creativity. The latter are the devices on the basis of which, for better and worse, humans end up questioning their condition, in all its dimensions, and potentially support or counter the very homeostatic mandate that permitted the questioning in the first place. Once again, the importance, efficiency, and even relative tyranny of genes are not in question. Their position in the order of things is.

LIFE ON EARTH

Beginning of Earth	+/- 4.5 billion years
Chemistry and protocells	4.0 to 3.8 billion years
First cells	3.8 to 3.7 billion years
Eukaryotic cells	2 billion years
Multicellular organisms	700 to 600 million years
Nervous systems	+/- 500 million years

3

VARIETIES OF HOMEOSTASIS

One of the first steps in the ritual otherwise known as the annual medical checkup is the measuring of blood pressure. All wise readers have had their blood pressure measured regularly and are familiar with the fact that there are ranges for the numbers the doctor announces, for the "diastolic" and "systolic" measurements. Some readers will even have had episodes of high or low blood pressure and been told to change their diets or take medications to bring the measurement into the acceptable range. Why so much fuss? Because there is a permissible range of variation for one's blood pressure and only limited fluctuations are allowed. The organism is expected to automatically regulate the process and avoid excessive deviations toward the lower and upper limits. But when that natural safety device fails, trouble ensues, sometimes immediately, if the degree of failure is high. When the failure persists, it has grave consequences for the future of the organism. What your doctor is seeking is evidence that one among several systems of your organism is or is not functioning as it should.

Homeostasis and life regulation are usually seen as synonymous. This is in keeping with the traditional concept of homeostasis, which refers to the ability, present in all living organisms, to continuously and automatically maintain their functional operations, chemical and general physiological, within a range of values compatible with survival. This narrow concept of homeostasis does not do justice to the complexity and reach of the phenomena to which the term refers.

It is certainly true that whether we consider unicellular life-forms or complex organisms, such as we are, very few aspects of an organism's operation escape the obligation to keep themselves in check. Accordingly, the mechanisms of homeostasis were first conceptualized as strictly automatic and pertained only to the state of an organism's internal environment. In keeping with this definition, the concept of homeostasis was often explained by analogy to a thermostat: upon reaching a previously set temperature, the device automatically commands itself to either suspend the ongoing operation—cooling or heating—or initiate it, as appropriate. The traditional definition, however, as well as the typical explanations it inspired, fail to capture the range of circumstances in which it can be applied to living systems. Let me explain why the traditional view is not sufficiently ample.

First, the homeostatic process strives for more than a mere steady state. Considered in retrospect, it is as if single cells or multicellular organisms were striving for a particular class of steady state conducive to flourishing. This is a natural upregulation that can be described as aiming at the future of the organism, an inclination to project itself in time by means of *optimized* life regulation and possible progeny. One might say that organisms want their health and then some.

Second, physiological operations rarely abide by thermostat-like set points. On the contrary, there are shades and grades of regulation; there are steps along scales that ultimately correspond to the greater or lesser perfection of the regulatory process. This process corresponds to what is commonly experienced as feelings, and the two issues are closely related: the former, the relative goodness or badness of a given life state, is the basis for the latter, that is, feelings. On this note, it is remarkable to consider that in general we do not need to visit our physician to discover if the fundamentals of our health are fine. Nor do we need a blood test for that purpose. Feelings provide us with a moment-to-moment perspective on the state of our health. Degrees of well-being or malaise are sentinels. Of course, feelings can miss the onset of several diseases, and emotional feelings can mask the ongoing, spontaneous homeostatic feelings and prevent them from delivering a clear message. More often than not, however, feelings tell us what we need to know. There is no reason why we should rely on feelings alone to take good care of ourselves. But it is important to point out the fundamental role of feelings and their practical value, no doubt the reason why they have been preserved in evolution.

Third, a comprehensive view of homeostasis must include the application of the concept to systems in which conscious and deliberative minds, individually and in social groups, can both interfere with automatic regulatory mechanisms *and* create new forms of life regulation that have the very same goal of basic automated homeostasis, that is, achieving viable, upregulated life states that tend to produce flourishing. *I see the effort of constructing human cultures as a manifestation of this variety of homeostasis.*

Fourth, whether one considers single-celled or multicellular organisms, the essence of homeostasis is the formidable enterprise of managing energy—procuring it, allocating it to critical jobs such as repair, defense, growth, and participation in the engendering and maintenance of progeny. This is a monumental endeavor for any organism, all the more so for human organisms given the complexity of their structure, organization, and environmental variety.

So large is the scale of the enterprise that its effects can begin at a low level of the physiology and manifest themselves at the higher levels of function, namely, cognition. For example, it is known that as ambient temperatures rise, not only do we need to adjust our internal physiology to losses of water and electrolytes, but we also function less well cognitively. That poor adjustment of internal physiology spells disease and death is no surprise. It is known that the number of deaths increases during prolonged heat waves, and heat waves also spell more murders and sectarian violence.[1] Students do significantly less well in exams, and civility is tied to the thermometer, too.[2] The relation between homeostasis and physiology holds for all levels of the living economy, from low to high. The clever cultural responses to heat waves, in all likelihood conceived in a shaded spot, began with fans and ended up with air-conditioning. So here is a good example of homeostatically driven technological developments.

The Distinct Varieties of Homeostasis

The traditional and narrow concept of homeostasis does not easily or usually conjure up the fact that nature evolved at least two distinct varieties of internal milieu control and that the single term "homeostasis" can refer to either variety or to both. As a result, the extraordinary significance of this evolutionary development is easily missed. The common usage of the term "homeostasis" refers to a nonconscious form of physiological control that operates automatically without subjectivity or deliberation on the part of the organism. Obviously, as we have seen in the case of bacteria, it can even operate well in organisms without a nervous system.

Indeed seeking food or drink when energy sources are depleted can be achieved by most organisms without any willful intervention on their part, and should food or drink not be available in the environment, most organisms cope with the problem automatically as well. Hormones will automatically break down stored sugars and deliver them to the blood so as to compensate for the immediate lack of energy sources. At the same time, the organism is automatically driven to intensify its search for energy sources. The primary result of such measures is survival, while the required solution—the ingestion of food—is not available. Similarly, when water balance is low, the kidneys automatically shut down or slow down their operation. This prevents or reduces diuresis and restores the level of hydration while the organism waits for better times. Hibernation is a natural coping strategy whenever temperature and energy availability fall short.[3]

For numerous living creatures, however, and certainly for humans, this narrow usage of the term "homeostasis" is inadequate. It is true that humans still make good use and greatly benefit from automatic controls: as noted, the value of glucose in the bloodstream can be automatically corrected to an optimal range by a set of complex operations that do not require any conscious interference on the part of the individual; the secretion of insulin from pancreatic cells, for example, adjusts the level of glucose; likewise, the amount of circulating water molecules can be automatically adjusted by diuresis. In humans and in numerous other species endowed with a complex nervous system, however, there is a supplementary mechanism that involves mental experiences that express a value. The key to the mechanism, as we have seen, is feelings. But as the terms "mental" and "experience" suggest, feelings, in the full sense implied here, could only come to pass once there were minds and the respective mental phenomena, and once minds could be made conscious and have experiences.[4]

Homeostasis Now

The sort of automated homeostasis that we find in bacteria, simple animals, and plants precedes the development of minds later to be imbued with feelings and consciousness. Such developments gave minds the possibility of deliberate interference with preset homeostatic mechanisms and even later allowed creative and intelligent invention to expand homeostasis into the sociocultural domain. Curiously, however, automated homeostasis, beginning with bacteria, included and in fact required sensing and responding abilities, the humble precursors to minds and consciousness. Sensing operates at the level of chemical molecules present in the membranes of bacteria and is found in plants as well. Plants can sense the presence of certain molecules in the soil— the tips of their roots are sensory organs, in fact—and they can act accordingly: they can grow in the direction of the terrain where the homeostatically required molecules are likely to be.[5]

The popular notion of homeostasis—if the reader can excuse the incongruity of having the words "popular" and "homeostasis" in the same sentence—conjures up the ideas of "equilibrium" and "balance." But we do not want equilibrium at all when we are dealing with life, because thermodynamically speaking equilibrium means zero thermal difference and death. (In the social sciences, the term "equilibrium" is more benign because it simply means the stability that results from comparable opposing forces.) We do not want to use "balance" either, because it conjures up stagnation and boredom! For years, I used to define "homeostasis" by saying that it corresponded not to a neutral state but to a state in which the operations of life felt as if they were upregulated to well-being. The forceful projection into the future was signified by the underlying feeling of well-being.

I recently encountered a kindred view in the formulations of John Torday, who also rejects the quasi-static view of homeostasis, the maintenance of status quo view. Instead, he embraces a view of homeostasis as a driver of evolution, a way into the creation of a protected cellular space within which catalytic cycles can do their job and literally come to life.[6]

The Roots of an Idea

We owe the idea behind homeostasis to the French physiologist Claude Bernard. In the last quarter of the nineteenth century, Bernard made a pathbreaking observation: living systems needed to maintain numerous variables of their internal milieu within narrow ranges so that life would continue.[7] In the absence of this tight control, the magic of life simply vanished. The essence of the internal milieu (*milieu intérieur* in the original) is a large number of interacting chemical processes. The typical chemical processes and their key molecules can be found in the bloodstream, in viscera, where they help accomplish metabolism, in endocrine glands such as the pancreas or the thyroid glands, and in certain regions and circuits of the nervous system where aspects of life regulation are coordinated—the hypothalamus is the prime example of such a region. These chemical processes enable the transformation of energy sources into energy itself by ensuring that water, nutrients, and oxygen are present as needed in living tissues. This is required for the cells that compose all body tissues and organs to maintain their individual lives. The organism, which is the integrated whole of all those living cells, tissues, organs, and systems, can only survive if the homeostatic limits are closely observed. Deviations from the required level of certain variables result in disease, and unless a more or less rapid correction occurs, the radical result is death. All living organisms are endowed with automated regulatory mechanisms. They are readily provided and come with a warranty signed by their genomes.

The actual term "homeostasis" was coined several decades after Claude Bernard by Walter Cannon, an American physiologist.[8] Cannon was also referring to living systems, and in inventing the name "homeostasis" for the process, he chose the Greek root *homeo-* (for "similar") and not *homo-* (for "same"), because he was thinking of systems engineered by nature, whose variables often exhibit workable ranges—hydration, blood glucose, blood sodium, temperature, and so forth. He was obviously not thinking about fixed set points, which are often present in systems engineered by humans, such as thermostats. The terms "allostasis" and "heterostasis," which are synonymous with "homeostasis," were introduced later with the valid purpose of calling attention to the issue of ranges, the fact that life regulation operates relative to ranges of values rather than set points.[9] The idea behind the more recent coinages, however, conforms to the idea implied by Bernard and named by Cannon with the original term. These newer terms have not entered common usage.[10]

I have greater sympathy for another term, "homeodynamics," coined by Miguel Aon and David Lloyd.[11] Homeodynamic systems, as is certainly the case with living systems, self-organize the operations when they lose stability. At those bifurcation points, they exhibit complex behaviors with emergent characteristics such as bistable switches, thresholds, waves, gradients, and dynamic molecular rearrangements.

Claude Bernard's proposal on the regulation of the internal milieu was so ahead of its time that it referred to not just animals but also plants. The mere title of his 1879 book is astonishing even today: *Leçons sur les phénomènes de la vie communs aux animaux et aux végétaux (Lectures on the Phenomena of Life Common to Animals and Plants).*

The kingdoms of plants and of animals have been traditionally conceived by their respective students as being far apart. But Claude Bernard understood that plants and animals have similar basic requirements. Plants are multicellular organisms that need water and nutrients as animals do; they have complicated metabolisms; they do not have neurons, muscles, or much overt movement, although there are a few brilliant exceptions, but they have circadian rhythms, and their homeostatic regulation uses some of the same molecules that our nervous system does—serotonin, dopamine, noradrenaline, and so forth. Plants are usually seen as immobile, but there is more movement in plants than meets the eye. I am not just referring to the Venus flytrap that briskly snaps its petals shut on adventurous insects. Or to the fact that certain flowers open up to the sunlight and modestly close up by nightfall. The very growth of roots or plant trunks constitutes movement generated by the sheer addition of actual physical elements. This can be easily demonstrated by speeding up the frames of a patiently obtained film document of plant growth.

Claude Bernard also understood that in both plants and animals, homeostasis benefited from symbiotic relationships. A good example: flowers whose scents attract bees, whose visits are required for their own fabrication of honey and accomplish the pollination that will offer the plant's seeds to the world.

We are discovering today that the scope of symbiotic arrangements is far larger than even Claude Bernard anticipated. It includes, for both animals and plants, organisms from yet another kingdom, that of bacteria, the vast and variegated domain of prokaryotes. Trillions of bacteria live in well-run housing projects within our organism, contributing goods to our lives and receiving lodging and feeding benefits in return.

4

FROM SINGLE CELLS TO NERVOUS SYSTEMS AND MINDS

Ever Since Bacterial Life

I will ask the reader to put human minds and brains aside, for a moment, and consider bacterial life instead. The goal is to see where and how life in single cells fits in the long history that leads to humanity. The exercise may sound a bit abstract, at first, because we are not used to seeing bacteria with the naked eye. But there is nothing abstract at all about microorganisms when you see them through a microscope and when you learn of the amazing things they accomplish.

There is no doubt that bacteria were the first life-forms and that they are with us today. But to say that they are still around because they were brave survivors would be a gross understatement. They happen to be the most numerous and varied inhabitants of Earth. Not only that, many species of bacteria are truly part of us humans. Many have become incorporated in larger cells of the human body, over the eons of evolution, and many bacteria live within each of us now, in largely harmonious symbiosis. There are more bacterial cells inside each human organism than there are human cells in that same organism. The difference is staggering, by a factor of 10. In the human gut alone, there are usually around 100 trillion bacteria, while in one entire human being there are only about 10 trillion cells, counting all types. The microbiologist Margaret McFall-Ngai is well justified when she says that "plants and animals are a patina on the microbial world."[1]

This huge success has its reasons. Bacteria are very intelligent creatures; that is the only way of saying it, even if their intelligence is not being guided by a mind with feelings and intentions and a conscious point of view. They can sense the conditions of their environment and react in ways advantageous to the continuation of their lives. Those reactions include elaborate social behaviors. They can communicate among themselves—no words, it is true, but the molecules with which they signal speak volumes. The computations they perform permit them to assess their situation and, accordingly, afford to live independently or gather together if need be. There is no nervous system inside these single-celled organisms and no mind in the sense that we have. Yet they have varieties of perception, memory, communication, and social governance. The functional operations that support all this "intelligence without a brain or mind" rely on chemical and electrical networks of the sort nervous systems eventually came to possess, advance, and explore later in evolution. In other words, later, much later in evolution, neurons and neuron circuits have come to make good use of older inventions that relied on molecular reactions and on components of the cell's body known as the cytoskeleton—literally, the skeleton inside the cell—and the membrane.

Historically, the world of bacteria—cells without nuclei, known as prokaryotes—was followed, about 2 billion years later, by the far more complicated world of nucleated cells, or eukaryotes. Multicellular organisms, or metazoans, came next, 700 to 600 million years ago. This long process of evolution and growth is full of examples of powerful cooperations, although accounts of this history usually give pride of place to competition. For example, bacterial cells cooperate with other cells so as to create the organelles of more complex cells. Mitochondria are examples of organelles, mini-organs within a cellular organism. Technically speaking, some of our own cells began by incorporating bacteria in their structure. Nucleated cells, in turn, cooperate to constitute tissues, and later these tissues cooperate to form organs and systems. The principle is always the same: organisms give up something in exchange for something that other organisms can offer them; in the long run, this will make their lives more efficient and survival more likely. What bacteria, or nucleated cells, or tissues, or organs give up, in general, is independence; what they get in return is access to the "commons," the goods that come from a cooperative arrangement in terms of indispensable nutrients or favorable general conditions, such as access to oxygen or advantages of climate. Consider this the next time you hear people deride international trade agreements as a bad idea. The notable biologist Lynn Margulis championed the idea of symbiosis in the construction of complex life at a time when the idea was hardly considered.[2]

The homeostatic imperative stands behind the processes of cooperation and also looms large behind the emergence of "general" systems, ubiquitously present throughout multicellular organisms. Without such "whole-body systems," the complex structures and functions of multicellular organisms would not be viable. The main examples of such developments are circulatory systems, endocrine systems (which are in charge of distributing hormones to tissues and organs), immune systems, and nervous systems.[3] Circulatory systems make possible the distribution of nutrient molecules and oxygen to every cell in a body. They distribute the molecules that result from digestion carried out in a gastrointestinal system and that need to be delivered throughout an organism. The cells cannot survive without the molecules, and the same applies to oxygen. Think of circulatory systems as the original Amazon business. Circulatory systems also accomplish something remarkable: they collect most of the waste products that result from metabolic exchanges and succeed in getting rid of them. Last, they extend two critical assistants of homeostasis: hormonal regulation and immunity. Still, nervous systems are the pinnacle of organism-wide, homeostatically dedicated systems, and I turn to them next.

Nervous Systems

When do nervous systems enter the evolutionary march? One good estimate is the Precambrian period, which ended 540 to 600 million years ago, an old vintage for certain but not that old when we compare it with the age of first life. Life, even multicellular life, managed quite well without nervous systems for about 3 billion years. We should reflect on this time line before we decide when perception, intelligence, sociality, and emotions made their first appearance on the world stage.

Seen in the perspective of today, as nervous systems entered the scene, they allowed complex, multicellular organisms to cope better with organism-wide homeostasis and thus permitted physical and functional expansions for such organisms. Nervous systems emerged as servants to the rest of the organisms—to bodies, more precisely—not the other way around. It is arguable that to a certain extent they remain servants today.

Nervous systems have several distinctive traits. The most important has to do with the cells that best define them: the neurons. They are *excitable*. This means that when a neuron becomes "active," it can produce an electric discharge that travels from the cell body to the axon—the fiber extension that emerges from the cell body—and, in turn, causes the release of molecules of a chemical—known as a neurotransmitter—at the point where one neuron contacts either another neuron or a muscle cell. At that point, which is known as a synapse, the released neurotransmitter activates the subsequent cell, be it another neuron or a muscle cell. Few other types of cells in the body manage a comparable feat, that is, combine an electrochemical process to make another cell spring into action. Neurons, muscle cells, and some sensory cells are the typical examples.[4] We can see this feat as a glorification of the bioelectrical signaling first accomplished modestly, in simple celled organisms such as bacteria.[5]

Another trait behind the uniqueness of nervous systems comes from the fact that nervous fibers, the axons that arise from the neuron cell body, terminate at almost every nook and cranny of the body—individual internal organs, blood vessels, muscles, skin, you name it. To do so, nervous fibers often travel long distances away from the centrally located parent cell body. The presence of that distant terminal envoy is properly reciprocated, however. In evolved nervous systems, a reciprocal set of nervous fibers travels in the opposite direction, from myriad body parts to the central component of the nervous system, which in the human case is the brain. The task of the fibers going from central nervous system to periphery is, in essence, the incitement of actions such as the secretion of a chemical molecule or the contraction of a muscle. Consider the extraordinary importance of those actions: by delivering a secreted chemical molecule to the periphery, the nervous system alters the operation of the tissues that receive it; by contracting a muscle, the nervous system generates movement.

Meanwhile, the fibers traveling in the opposite direction, from the organism's interior toward the brain, perform an operation known as interoception (or visceroception because their job has so much to do with what is going on in the viscera). What is the purpose of such an operation? Surveillance over the state of life, that is what it is for, in a nutshell, a massive snooping and reporting job whose goal is to let the brain know what is going on elsewhere in the body so that it can intervene when needed and appropriate.[6]

Some details need to be noted in this regard. First, the neural surveillance job of interoception is heir to an earlier and more primitive system that permits chemical molecules traveling in the blood to act *directly* on both central and peripheral nerve structures. This ancient route of chemical interoception informs the nervous system about what is going on in the body proper. Clearly, this ancient route is reciprocated, in the sense that chemical molecules originating in the nervous system enter the bloodstream and can influence aspects of metabolism.

Second, in conscious creatures such as we are, the first tier of visceroception signals is delivered below the level of consciousness, and the corrective responses that the brain produces on the basis of unconscious surveillance are, for the most part, not consciously deliberated either. Eventually, as we shall see, the surveillance job yields conscious feelings and enters the subjective mind. It is only beyond that point of functional capability that the responses can be influenced by conscious deliberation while still benefiting from the nonconscious process.

Third, massive surveillance of organism functions, an advantageous development for adequate homeostasis in complex multicellular organisms, is the natural precursor to the "Big Data" surveillance technologies humans are so shamelessly proud of inventing. The surveillance is useful on two counts: straightforward information on the state of the body and, relatedly, anticipation and prediction of future states.[7] Here is another example of the strange order in which biological phenomena emerged in the history of life.

In brief, the brain acts on the body by delivering specific chemical molecules either to a particular body region or to the circulating blood, which subsequently routes the molecules to varied body regions. The brain can also even more literally *act* on the body by activating its muscles, both the muscles that we move when we *want* to—we can decide to walk or run or pick up a cup of coffee—and the muscles that are brought into action as needed, through no will of our own. For example, if you are dehydrated and your blood pressure is dropping, the brain orders the smooth muscles in the walls of your blood vessels to contract and thus increase the blood pressure. Likewise, the smooth muscles in your gastrointestinal system march to their own drum and produce digestion and nutrient absorption with little or no interference from you. The brain executes homeostatic compensations on behalf of the whole organism, and "we" benefit from them, effortlessly. A slightly more complicated level of involuntary movement is engaged when we spontaneously smile, laugh, yawn, breathe, or have hiccups, involuntary actions that require striated muscles. The heart is a cleverly and involuntarily controlled striated muscle.

The beginning of nervous systems was not this complicated; in fact, it was quite modest. It literally consisted of nerve nets, a reticulum, or network, of wires (the term derives from the Latin for "net," *rete*). The nerve nets of yore actually resemble the structure of the "reticular formations" that we can still find, today, in the spinal cord and the brain stem of so many species, humans included. In those simple nervous systems, there is not a sharp distinction between "central" and "peripheral" components. They consist of a wiring of neurons that crisscross the body.[8]

Nerve nets first appear in species such as cnidarians, in the Precambrian period. Their "nerves" arise from the body's external cellular layer, the ectoderm, and their distribution helps accomplish in a simple manner some of the main functions that complex nervous systems came to accomplish, far later in evolution, and still do. The more superficial nerves serve elementary perceptual purposes as they are stimulated from the outside of the organism. They sense the organism's surround. Other nerves can be used to move the organism, in response, for example, to an external stimulus. This is locomotion made simple—swimming, actually, in the case of hydras. Still another group of nerves can take care of regulating the organism's visceral environment. In the case of hydras, which are dominated by their gastrointestinal systems, the nerve nets take care of the entire sequence of gastrointestinal operations: ingestion of water with nutrients, the digestive business, and the excretion of wastes. The secret to these operations is peristalsis. The nerve nets deliver the goods by activating sequential muscular contractions along the digestive tube and producing peristaltic waves, not that different from our own, come to think of it. Curiously, sponges, which were once thought not to have nervous systems at all, display an even simpler variety of devices that control the caliber of their tubular cavities and thus allow, once again, admission of water with nutrients and expulsion of water with wastes. In other words, sponges distend and open up, or contract and shut themselves off. When they contract, they "cough" or "belch," as it were.

How intriguing, in this context, that the enteric nervous system—the complicated mesh of nerves that is present in our gastrointestinal tracts—so resembles old nerve net structures. This is one of the reasons why I suspect the enteric nervous system was really the "first" brain, not the "second" as it is popularly known.

It probably took millions of additional years—through the Cambrian explosion and beyond—to develop the more complex nervous systems of countless species, eventually culminating in the hugely complicated nervous systems of primates, especially humans. While the nerve nets of hydras can coordinate numerous operations and harmonize homeostatic needs with the conditions of the external environment, their capacities are limited. They can *sense* the presence of certain stimuli in the environment so that some convenient response can be triggered. The sensing capacities of hydras are, to be generous, a poor man's version of touch. In the most benign account we can offer, nerve nets accomplish very basic perception. The nets also do visceral regulation, a sort of beginner's autonomic nervous system, they run locomotion, and they coordinate all of these functions.

It is just as important to understand what nerve nets cannot accomplish. Their sensing permits useful and nearly instantaneous responses. The neurons that actually sense and act are modified by their activity and thus learn something regarding the events in which they are involved, but little knowledge is retained from the day-to-day existence of the respective organisms, a way of saying that their memory is limited. Their perception is also simple. The nerve net design is simple, and there is nothing in it that would allow much mapping of the constitutive aspects of a stimulus—a shape or a texture—or of its consequences for the organism. The structure of nerve nets would not allow them to represent the configurational pattern of an object touched. They lack mapping ability, and that also means that nerve nets cannot generate the images that eventually come to constitute the minds that complex nervous systems create so prolifically. The absence of mapping and image-making capabilities entails other fatal consequences: consciousness cannot arise in the absence of minds, and the same applies, even more fundamentally, to the very special class of processes called feelings, which are constituted by images closely interwoven with body operations. In other words, in my perspective and in the ample and technical sense of the terms, consciousness and feeling depend on the existence of minds. Evolution had to wait for more sophisticated nervous devices so that brains would be capable of fine multisensory perceptions based on the mapping of numerous component features. Only then, as I see it, was the way clear for the creation of images and for the construction of minds.[9]

Why was having images so important? What did having images really accomplish? The presence of images meant that each organism could create *internal representations* based on its ongoing sensory descriptions of *both* external and internal events. Those representations, generated within the organism's nervous system but with the cooperation of the body proper, made a world of difference to the particular organism in which such processes took place. Those representations, which were accessible *only* to the particular organism, were able, for example, to guide movement of a limb or of the whole with precision. Image-guided movements—guided by visual, sound, or tactile images—were more beneficial for the organism, more likely to produce advantageous results. Homeostasis was improved accordingly and with it survival.

In brief, images were advantageous even if an organism were not conscious of the images formed within it. The organism would not yet be capable of subjectivity and would be unable to inspect the images in its own mind, but still the images could automatically guide the execution of a movement; the movement would be more precise in terms of its target and succeed rather than fail.

As nervous systems developed, they acquired an elaborate network of peripheral probes—the peripheral nerves that are distributed to every parcel of the body's interior and to its entire surface, as well as to specialized sensory devices that enable seeing, hearing, touching, smelling, and tasting.

Nervous systems also acquired an elaborate collection of aggregated central processors in the central nervous system, conventionally called the brain.[10] The latter includes (1) the spinal cord; (2) the brain stem and the closely related hypothalamus; (3) the cerebellum; (4) a number of large nuclei located above brain-stem level—in the thalamus, basal ganglia, and basal forebrain; and (5) the cerebral cortex, the most modern and sophisticated component of the system. These central processors manage learning and memory storage of signals of every possible sort and also manage the integration of these signals; they coordinate the execution of complex responses to inner states and incoming stimuli—a critical operation that includes drives, motivations, and emotions proper; and they manage the process of image manipulation that we otherwise know as thinking, imagining, reasoning, and decision making. Last, they manage the conversion of images and of their sequences into symbols and eventually into languages—coded sounds and gestures whose combinations can signify any object, quality, or action, and whose linkage is governed by a set of rules called grammar. Equipped with language, organisms can generate continuous translations of nonverbal to verbal items and build dual-track narratives of such items.

Of special note are certain divisional arrangements of main functions that are organized and coordinated by different brain components. For example, several nuclei in the brain stem, hypothalamus, and telencephalon are in charge of producing the behaviors to which I referred above, known as drives, motivations, and emotions with which the brain responds to a variety of internal and external conditions with preset programs of actions (e.g., secretion of certain molecules, actual movements, and so forth).

Another important divisional arrangement pertains to the execution of movement and the learning of movement sequences. The cerebellum, basal ganglia, and sensorimotor cortices are the main players here. There are also principal divisions concerned with learning and recall of image-based facts and events—the hippocampus and the cerebral cortex, whose circuitry feeds into it and is reciprocated by it, are the star players. Yet another division permits the construction of verbal language translations for all the nonverbal images that the brain generates and streams in narratives.

It is to nervous systems so richly equipped and capable that the ability to feel is eventually conferred, as a coveted prize for achievements in mapping and imaging of internal states. And it is to such mapmaking and image-making organisms that the dubious prize of consciousness will also be attributed.

The glories of the human mind, the ability to memorize extensively, feel resonantly, translate any image and relationship of images in verbal codes, and generate all sorts of intelligent responses, can only come late in this story of numerous and parallel developments in nervous systems.

It is fair to say that a lot is known about the entire nervous system and that the main function of many of the components I have just enumerated has been generally elucidated. But it is also clear that numerous details of the operation of microscopic and macroscopic neural circuits are not known and that the functional integration of the anatomical components has not been fully conceptualized. For example, because neurons can be described as active or not, their operation lends itself to a description in terms of Boolean algebra, zeros or ones. This is a core belief behind the idea of brains as computers.[11] But microcircuit neural operations reveal unexpected complexities that undermine that simple view. For example, under certain circumstances, neurons can communicate to other neurons directly without using synapses, and neurons and the supporting glia also interact abundantly.[12] The result of these atypical contacts is a modulation of the neuron circuits. Their operations can no longer conform to the simple on/off schema and cannot be accounted for by the simple digital design. Moreover, the relation between brain tissue and the body that the brain is inserted in has not been fully understood. Yet that relation is key to providing a full account of how we feel, of how consciousness is constructed, and of how our minds engage in intelligent creations, the aspects of brain function that are most significant to explain our humanity.

In the effort to address these questions, I believe it is important to place the human nervous system in a suitable historical perspective. That perspective requires an acknowledgment of the following facts:

1. that the emergence of the nervous system was an indispensable enabler of life in elaborate multicellular organisms; the nervous system has been a servant of whole-organism homeostasis, although its cells also depend on that same homeostasis process for its own survival; this integrated mutuality is most often overlooked in discussions of behavior and cognition;

2. that the nervous system is part of the organism it serves, specifically a part of its body, and that it holds close interactions with that body;

that these interactions are of an entirely different nature from those that the nervous system holds with the environment that surrounds the organism; the particularity of this privileged relationship also tends to be overlooked; I will say more on this critical issue in part II;

3. that the extraordinary emergence of the nervous system opened the way for neurally mediated homeostasis—an addition to the chemical/visceral variety; later, after the development of conscious minds capable of feeling and creative intelligence, the way was open for the creation, in the social and cultural space, of complex responses whose existence began as homeostatically inspired but later transcended homeostatic needs and gained considerable autonomy; therein the beginning but not the middle or the end of our cultural lives; even at the highest levels of sociocultural creation, there are vestiges of simple life-related processes present in the most humble exemplars of living organisms, namely, bacteria;

4. that several complex functions of the higher nervous system have their functional roots in simpler operations of the lower devices of the system itself; for this reason, for example, it has not been productive to first look for the grounding of feeling and consciousness in the operations of the cerebral cortex; instead, as discussed in part II, the operation of brain-stem nuclei and of the peripheral nervous system offers better opportunities to identify precursors to feeling and consciousness.

The Living Body and the Mind

We are commonly given accounts of mental life—of perceptions, feelings, ideas, of the memories with which perceptions and ideas can be recorded, of imagination and reasoning, of the words used to translate internal narratives, inventions, and so forth—as if they were the exclusive product of brains. The nervous system is often the hero of these accounts, both an oversimplification and a misunderstanding. It is as if the body were a mere bystander, a support for the nervous system, the vat in which the brain fits.

That the nervous system is the enabler of our mental life is not in doubt. What is missing from the traditional neuro-centric, brain-centric, and even cerebral-cortex-centric accounts is the fact that nervous systems began their existence as assistants to the body, as coordinators of the life process in bodies complex and diversified enough that the functional articulation of tissues, organs, and systems as well as their relation to the environment required a dedicated system to accomplish the coordination. Nervous systems were the means to achieve that coordination and thus became an indispensable feature of complex multicellular life.

A more sensible account of our mental life is that both its simple aspects and its extraordinary achievements are partial by-products of a nervous system that delivers, at a very complex physiological level, what simpler life-forms have long been delivering without nervous systems: homeostatic regulation. On the way to accomplishing the principal task of making life possible in a complex body, nervous systems developed strategies, mechanisms, and abilities that not only took care of vital homeostatic needs but also produced many other results. Those other results were either not immediately necessary for life regulation or less clearly related to it. Minds depend on the presence of nervous systems charged with helping run life efficiently, in their respective bodies, and on a host of interactions of nervous systems and bodies. "No body, never mind." Our organism contains a body, a nervous system, *and* a mind that derives from both.

Minds can soar above their fundamental mission and yield products that, at first glance, are not homeostasis related.

The story of the relations between bodies and nervous systems needs to be revised. The body, about which we are often casual if not dismissive when we talk about the lofty mind, is part of a massively complex organism made up of cooperative systems, which are made up of cooperative organs, which are made up of cooperative cells, which are made up of cooperative molecules, which are made up of cooperative atoms built from cooperative particles.

One of the most distinctive features of organisms is, in fact, the extraordinary degree of cooperation exhibited by their constituent elements, along with the resulting extraordinary complexity. Just as life emerged from particular relationships between cellular elements, so, too, the increasing complexity of organisms results in new functions. The emergent functions and qualities cannot be explained by simply examining the individual constituents. In brief, complexity is hallmarked by the appearance of functional emergences as one moves from smaller to bigger chunks of the overall structure. The prime example is the distinctive emergence of life itself, in cellular elements. Another prime example of cooperation, about which more later, is the emergence of subjective mental states.

There is more to organism life than the sum total of the lives of each participating cell. There is an overall life of the organism, a *global* life, as it were, resulting from the high-dimension integration of the contributing lives within it. Organism life transcends the lives of its cells, draws on them, and reciprocates the favor by sustaining them. That integration of real "lives" is what makes a whole organism alive in precisely the same sense that a current complex computer network is not. Organism life means that *each component cell* still needs to use and is capable of using its elaborate microscopic components to transform nutrients captured from its environment into energy; it does so under elaborate rules of homeostatic regulation and under the homeostatic imperative of preserving itself against all odds and carrying on. But the extraordinary complexity of a living organism, the human variety being the best example, could only have come to be with the help of the supporting, coordinating, and controlling devices of the nervous system. All these systems are entirely part of the body that they serve. In and of themselves, they, too, are made up of living cells, like all the rest. Their cells also require regular nourishment to preserve their integrity, and they, too, are at risk of disease and death, just like any other cell in the body.

The order of appearance of organs, systems, and functions in living organisms is critical for the understanding of how some of those functions emerged and came to operate. Nowhere is this more apparent than in the need to consider the *precedences* of parts and functions in the history of nervous systems, most notably the human nervous system and its magnificent products: mind and culture. There is an order to the emergence of things, and it is strange or not that strange depending on one's perspective.

PART II

ASSEMBLING

THE CULTURAL MIND

5

THE ORIGIN OF MINDS

The Momentous Transition

How does one get from the deceptively simple life of nearly 4 billion years ago to the life of the past 50,000 or so years, the life that harbors human cultural minds? What can we say about the trajectory and about the instruments it used? To say that natural selection and genetics are a key to the transformation is entirely true but not enough. We need to acknowledge the presence of the homeostatic imperative— put to beneficial use or not—as a factor in the selective pressures. We need to acknowledge the fact that there was neither a single line of evolution nor a simple progression of complexity and efficiency of organisms, that there were ups and downs and even extinctions. We need to note that a partnership of nervous systems and bodies was required to generate human minds and that minds occurred not to isolated organisms but to organisms that were part of a social setting. Last, we need to note the enrichment of minds by feelings and subjectivity, image-based memory, and the ability to enchain images in narratives that probably began as nonverbal, film-like sequences but eventually, after the emergence of verbal languages, combined verbal and nonverbal elements. The enrichment came to include the ability to invent and produce intelligent creations, a process I like to call "creative intelligence" and that is a step up from the smarts that enable numerous living organisms, including humans, to behave efficiently, quickly, and winningly in everyday life. Creative intelligence was the means by which mental images and behaviors were intentionally combined to provide novel solutions for the problems that humans diagnosed and to construct new worlds for the opportunities humans envisioned.

I address these issues in this and in the next four chapters, beginning with the origins and making of minds and ending with the mental components that originally enabled creative intelligence, namely, feeling and subjectivity. The goal here is not to treat the psychology and biology of such abilities comprehensively but rather to sketch their nature and acknowledge their roles as instruments of the human cultural mind.

Minded Life

In the beginning, it was just sensing and responding in a one-celled organism capable of some movement of its whole body. To imagine what sensing and responding were like, one needs to picture pores in the membrane envelope of the cell and realize that when certain molecules were present at these pores, they served as chemical signals to other cells and received signals from other cells and from the environment. Imagine something like emitting a scent and smelling a scent. Sensing and responding were first made of this: exhibit a signal that signifies a living presence and be signaled in return by comparably equipped creatures. The signals resembled an irritant and produced a corresponding irritability. There were no "eyes" or "ears," although you can say that the sensing molecules behaved as if there were.[1] Smells and tastes would be a closer analogy, but they were not that either. There was nothing "mental" about this process. Inside the cell, there were no representations that *resembled* either the external world or the world inside, nothing that one might call an image, let alone a mind or consciousness. There was merely a beginning of the sort of perceptual process that, over time, once nervous systems came onto the scene, would indeed lead to analog representations of the world surrounding nervous systems and serve as the basis for minds and eventually subjectivity. The march toward minds began with elementary sensing and responding, and sensing and responding are still at work today in the world of the bacteria that live inside our organisms and in every animal, plant, water, and soil, and even the depth of the earth. In bacteria, sensing and responding signal the presence of others, and even help with an estimate of how many others there are in the vicinity. But plain sensing and responding do not require the properties of mind and the properties that flow from mind. Bacteria and many other single-celled organisms are not mindful or conscious other than figuratively. And yet sensing and responding are contributors to what eventually becomes more complex perceptions and mind. If we are to account for the latter, we need to acknowledge and understand the former and glean the chain that links them. The sensing and responding level of perception precedes minds, historically speaking, and is also present in minded organisms *now*. In most normal situations, our own minds respond to material that is being sensed and engender further responses, in the form of mental representations and mentally directed actions. We only suspend basic sensing and responding when we are under anesthesia and during sleep, and even then not completely.[2]

Eventually, there came organisms with many cells. Their movements were finer. Internal organs began to appear and became more differentiated. A significant novelty was the refinement of whole-body systems and the appearance of new ones. Instead of single-function organs—guts, hearts, lungs—general systems covered the territory. Unlike individual cells, which looked largely after their own business, general systems were made from many cells and looked after the business of *all* other cells inside a multicellular organism. They were dedicated, for example, to the circulation of fluids such as lymph and blood; to producing internal and eventually external movements; and to the global coordination of organism operations. The coordination was provided by the endocrine system via chemical molecules known as hormones and by the immune system, which ensured inflammatory responses and immunity. The master global coordinators followed suit. They were, of course, the nervous systems.

Jump to a few billion years later and organisms were now very complicated, and so were the nervous systems that helped them fend for themselves and stay alive. Nervous systems had become capable of sensing different parts of the environment—physical objects, other living creatures—and responding with appropriate movements of sophisticated limbs and of the whole body: grab, kick, destroy, run away from, touch gently, have sex. Nervous systems and the organisms they served worked in full cooperation.

At some point, long after nervous systems were able to respond to many features of the objects and movements that they sensed, both outside *and* inside their own organism, there began the ability to *map* the objects and events being sensed. This meant that rather than merely helping detect stimuli and respond suitably, nervous systems literally began drawing maps of the configurations of objects and events in space, using the activity of nerve cells in a layout of neural circuits. To catch a glimpse of how this worked, imagine the neurons wired in circuits and laid out on a flat board, where every point of the surface corresponds to a neuron. Then imagine that when a neuron in the circuitry becomes active, it lights up, something akin to making a dot on a board using a marker. The ordered, gradual addition of many such dots generates lines that can link up or intersect and create a map. Let me give the simplest of examples. When the brain makes a map of an object with the shape of an X, it activates neurons along two linear rows that intersect at the appropriate point and angle. The result is a neural map of an X. The lines in brain maps represent the configuration of an object, its sensory features or movements or location in space. The representation does not need to be "photographic," although it can be. It is essential, however, that it preserve the internal relations among the parts of an entity such as angles between components, superpositions, and so forth.[3]

Now stretch your imagination and think of maps not just of shapes or spatial locations but also of sounds as they occur in space, soft or rough, loud or faint, close or far away, and also think of maps built from touch, or smell or taste. Stretch the imagination a bit more and think of maps built from the "objects" and "events" that occur within the organism, that is, the viscera and their operations. Finally, the depictions produced by this web of nervous activity, the maps, are none other than the contents of what we experience as images in our minds. The maps of each sensory modality are the basis for the integration that makes images possible, and those images as they flow in time *are* the constituents of minds. They are a transformative step in the existence of complex living organisms, a fine consequence of the body–nervous system cooperation I have been addressing. Human cultures would never have come to pass without this step.

The Big Conquest

The ability to generate images opened the way for organisms to *represent the world around them*, a world that included every possible kind of object *and* other whole organisms; and, just as important, it allowed organisms to *represent the world inside each of them*. Before the emergence of mapping and images and minds, organisms could acknowledge the presence of other organisms and of external objects and respond accordingly. They could *detect* a chemical molecule or a mechanical stimulus, but the detection process did not include description of the *configuration* of the object that emitted the molecule or shoved the organism. Organisms could sense the presence of another organism because a *part* of that other organism had come into contact. They could also reciprocate the favor and be sensed. But the arrival of mapmaking and images provided a novel possibility: organisms could now produce a *private representation of the universe surrounding their nervous systems.* This is the formal beginning, in living tissues, of signs and symbols that "depict" and "resemble" the objects and events that the sensory channels of vision and hearing or touch manage to detect and describe.

The "surround" of a nervous system is extraordinarily rich. It literally is far more than meets the eye. It includes the world external to the organism—the only surround that is commonly and regrettably thought of, by scientists and laypersons alike, in discussions of this sort, that is, the objects and events in the environment surrounding the *whole* organism. But the "surround" of the nervous system also includes the world *within* the organism in question, and this part of the surround is commonly ignored to the peril of realistic conceptions of general physiology and of cognition in particular.

I believe that the possibility of representing the entire surround of nervous systems within that same nervous system, the availability of these nonpublic, internal manifestations, set the evolution of organisms on a new course. These are the "ghosts" that living organisms lacked, in all probability the ghosts Friedrich Nietzsche envisioned when he thought of humans as "hybrids of plants and of ghosts." Eventually, side by side, nervous systems working with the rest of their body would create internal images of the universe around the organism in parallel with images of the organism's interior. We were entering, at long last, quietly and modestly, the era of the mind, the era whose essence is still with us. We could now string together images in such a way that the images could narrate, to the organism, *both* internal events and events external to it.

On this account, the steps that must have followed in evolution are fairly clear. First, using images made from the oldest components of the organism's interior—the processes of metabolic chemistry largely carried out in viscera and in the blood circulation and the movements they generated—nature gradually fashioned feelings. Second, using images from a less ancient component of the interior—the skeletal frame and the muscles attached to it—nature generated a representation of the encasement of each life, a literal representation of the house inhabited by each life. The eventual combination of these two sets of representations opened the way for consciousness. Third, using the same image-making devices and an inherent power of images— the power to stand for and symbolize something else—nature developed verbal languages.

Images Require Nervous Systems

Elaborate life processes can well exist without nervous systems, but elaborate multicellular organisms *need* nervous systems to run their lives. Nervous systems play important roles everywhere in the management of organisms. Here are some examples: they coordinate movement—internally, in viscera, and externally through the use of limbs; they coordinate the internal production and delivery of chemical molecules required to maintain living conditions in partnership with the endocrine system; they coordinate the overall behavior of organisms relative to the natural dark and light cycles, and they run the related operations of sleep and wakefulness and the requisite changes in metabolism; they coordinate the maintenance of body temperature suitable to the continuation of life; last, but definitely not least, nervous systems make maps, which, as images, are the main ingredients of minds.

The existence of images was not possible before nervous systems grew in complexity. The world of sponges or of cnidarians such as hydras was enriched by the gift of a simple nervous system, but image making is unlikely to have been among its capabilities.[4] We can only guess, but minds that resemble ours in some elementary way belong to far more elaborate creatures whose nervous systems and behaviors had developed great complexity. In all likelihood, they are present in insects, for example, and probably in all or most vertebrates. Birds clearly have minds, and by the time we get to mammals, their minds must have enough resemblance to ours that we treat some of the respective creatures with the natural assumption that they understand not just what we do but often how we feel and sometimes how we think. Just consider chimpanzees, dogs and cats, elephants and dolphins, wolves. The fact that they lack verbal language, that their memory capacity and intellect are arguably less prodigious than ours, and that, consequently, they have not generated cultural artifacts comparable to those of humans is obvious. Still, the kinship and resemblances are overwhelming, and they are important to help us understand ourselves and how we came to be what and who we are.

Nervous systems are rich in mapmaking devices. The eye and the ear map varied features of the visual world and of the world of sound, respectively, in the retina, in the inner ear, and continue doing so in the central nervous system structures that follow them in a sequence that digs deep within the cerebral cortex. When we touch an object with our fingers, the nerve terminals distributed in the skin map the varied features of an object: the overall geometry, the texture, the temperature, and so forth. Taste and smell are two other channels for the mapping of the outside world. Advanced nervous systems such as ours fabricate, and abundantly so, *images of the outside world* and *images of the world inside the respective organisms*. In turn, the images of the world inside are of two very distinct sorts, relative to their source and contents: *the old and the not so old internal worlds.*

Images of the World Outside Our Organism

The images of the outside world originate in sensory probes located in the surface of the organism collecting information about all manner of details in the physical structure of the world around us. The traditional five senses—vision, hearing, touch, taste, and olfaction—have specialized organs in charge of collecting the information (but see note 5, below, for the vestibular sense that is closely related to hearing). Four of the five specialized organs—for vision, hearing, taste, and smell—are located in the head and are relatively close to one another. The organs of smell and taste are distributed in small patches of mucosae, a variant of skin that is naturally kept moist and is protected from direct sunlight; it lines the nasal and oral cavities. The specialized organ of touch is distributed over the entire skin surface and the mucosae. Curiously, there are taste receptors in the gut, no doubt a remnant from the days when the gut and its nervous systems were the only game in town.[5]

Each sensory probe is devoted to sampling and describing specific aspects of what the outside world is like in terms of its countless features. None of the five senses alone produces a comprehensive description of the outside world, although our brains eventually integrate the partial contributions of each sense into an overall description of an object or event. The result of this integration approximates a "whole" object description. On its basis, it is possible to generate a reasonably comprehensive image of an object or event. This is unlikely to be a "complete" description, but it is, certainly for us, a rich sampling of features, and it is all we have got, anyway, given the nature of the reality around us and the design of the senses. Fortunately, all of us are

immersed in this same incompletely sampled "reality," and we all suffer from comparable "imaging" limitations. It is a level playing field
among humans, and in good part we share that level playing field with
other species.[6]

The specialization of the nerve terminals of each sense is truly
astounding, each matched over evolutionary time to specific features and traits of the universe around. Chemical and electrochemical signals are the means by which the sensory terminals transmit
information from the outside in, across peripheral nerve pathways
and structures of the lower components of the central nervous system such as nerve ganglia, spinal cord nuclei, and lower brain-stem
nuclei. However, the critical function on which image making depends
is mapping, often macroscopic mapping, the ability to plot the different data arising from sampling the outside world in some sort of car-
tography, a space within which the brain can plot patterns of activity
and the spatial relationship of the active elements in the pattern. This
is how the brain maps a face that you see, or the contour of a sound, or
the shape of the object you are touching.

Images of the World Internal to Our Organism

There are two kinds of worlds inside our organisms. Let us call them
the old interior world and the not so old. The *old* internal world is
concerned with basic homeostasis. This is the very first and oldest
interior world. In a multicellular organism, this is the interior world
of metabolism complete with its related chemistries, of viscera such
as hearts, lungs, guts, and skin, and of the smooth muscles that can be
found everywhere in the organism where they help build the walls of
blood vessels and the capsules of organs. Smooth muscles are also, in
and of themselves, visceral elements.

The images of the internal world are the ones that we describe with such terms as "well-being," "fatigue," or "malaise"; "pain" and "plea-sure"; "palpitations," "heartburn," or "colics." They are of a special kind because we do not "picture" the old interior world in quite the same way that we picture objects out in the world. There is less detail, to be sure, although we can mentally illustrate the changing geom-etries of viscera in the idiom of visceral sensations—the tightening of the pharynx and larynx that occurs when we are in fear, or the tightening of the airway and the gasping for air characteristic of an asthma attack; likewise, for the effects of certain molecules on varied components of the body, which often include motor reactions such as tremors. These images of the old internal world are none other than core components of *feelings*.

Side by side with the old internal world there is also a *newer* inter-nal world. This one is dominated by our bony skeleton and by the muscles attached to it, the skeletal muscles. The skeletal muscles are also known as "striated" or "voluntary"; this helps distinguish them from the "smooth" or "autonomic" variety, which is purely visceral and not under our willful control. We use skeletal muscles to move about, manipulate objects, speak, write, dance, play music, and operate machinery.

The overall body frame, inside of which part of the old visceral world is located, is the scaffolding over which the old world of the skin is literally draped. Note that the skin is the largest of our viscera. Note also that the overall body frame is the setting in which our *sensory portals* are placed, encrusted like so many jewels in a piece of compli-cated jewelry.

With the term "sensory portals," I am referring to the regions of the body frame where the sensory probes are implanted and to the sensory probes themselves. Four of the principal sensory probes are well circumscribed: the eye sockets, the musculature controlling the eyes, and the machinery inside the eyes; our ears, including the tym-panic cavity and the tympanic membrane and the adjoining vestible in charge of sensing our position in space, that is, our balance; the nose and its olfactory mucosae; the taste buds in our tongue. As for the fifth portal, the skin with which we can touch any object and appreciate textures, it is distributed all over the body, although its perceptual abilities are distributed unevenly, since they are concentrated predominantly in the hands, mouth, mammillary, and genital regions.

The reason why I devote so much attention to the notion of the sensory portal has to do with its role in the generation of perspective. Let me explain. Our vision, for example, is the result of an enchainment of processes that begin in the retina and continue across several stations of the visual system—for example, the optic nerves, the superior geniculate nuclei and superior colliculi, the primary and secondary visual cortices. But to produce vision, we also need to engage in the *acts of looking and seeing,* and those acts are accomplished by *other* structures of the body (varied muscular groups) and of the nervous system (motor control regions), *separate from the visual system stations.* Those other structures are located at the visual sensory portal.

What does the visual sensory portal consist of? The eye socket; the musculature in our eyelids and around the eyes with which we can frown and concentrate gaze; the lens with which we adjust the visual focus; the diaphragm with which we control the amount of light; the muscles with which we move the eyes. All these structures and their respective actions are well coordinated with the primary visual process but are not part of it. They play an obviously practical role; they are assistants, so to speak. They also play a somewhat loftier and unintended role that I will address later when we turn to consciousness.

The *old* internal world is a world of fluctuating life regulation. It can operate well or not so well, but how well it works is critical to our lives and minds. Accordingly, the imaging of the old internal world in action—the state of viscera, the consequences of chemistries—must reflect the goodness or badness of the state of that interior universe. The organism needs to be affected by such images. It cannot afford to be indifferent to them, because survival depends on the information that such images reflect regarding life. Everything in this old internal world is qualified, good, bad, or in between. This is a world of *valence*.

The *new* internal world is a world dominated by the body frame, by the location and state of the sensory portals within that frame, and by the voluntary musculature. The sensory portals sit and wait within the body frame and contribute importantly to the information generated by the maps of the outside world. They clearly indicate to the organism's mind the *locations*, within the organism, of the sources of images currently being generated. This is necessary for the construction of an overall organism image, which, as we shall see, is a critical step in the generation of subjectivity.

The *new* internal world also generates valence because its living flesh does not escape the vagaries of homeostasis. But the vulnerabilities of the new internal world are smaller than those of the old. The skeleton and the skeletal musculature form a protective carapace. It sturdily envelops the tender old world of chemistries and viscera. The new internal world stands in relation to the old internal world as an engineered exoskeleton stands in relation to our real skeleton.

6

EXPANDING MINDS

The Hidden Orchestra

The poet Fernando Pessoa saw his soul as a hidden orchestra. "I do not know which instruments grind and play away inside of me, strings and harps, timbales and drums," he wrote in *The Book of Disquiet*.[1] He could only recognize himself as a symphony. His is an especially apt intuition, because the constructions that inhabit our minds can well be imagined as ephemeral musical performances, played by several hidden orchestras, inside the organisms to which they belong. Pessoa did not express puzzlement as to who would be playing all those hidden instruments. Perhaps he saw himself multiplied and doing all the playing, a bit like Oscar Levant in *An American in Paris*, not a surprising turn for a poet who invented so many aliases.[2] But we may well ask, who are the players in these imaginary orchestras, exactly? And here is the answer: *the objects and events in the world around our organisms, actually present or recalled from memory, and the objects and events in the world inside.*

What about the instruments? Pessoa could not identify the instruments that he could hear so well, but we can do that for him. There are two groups of instruments in the Pessoa orchestra. First, the main *sensory devices* with which the world around and inside an organism interacts with the nervous system. Second, the devices that continuously respond emotively to the mental presence of any object or event. The emotive response consists of altering the course of life within the old interior of organisms. The devices are known as drives, motivations, and emotions.

The varied players—objects and events, currently present or recalled from memory—do not pluck the strings of any violins or cellos and do not press the keys of countless pianos, but the metaphor captures the situation. Objects and events do "play," in the sense that they, as distinct entities within the organism's mind, can act on certain neural structures of the organism, "affect" their state, and change those other structures for a passing moment. Over the "playing time," their actions result in a certain kind of music, the music of our thoughts and feelings and of the meanings that emerge from the inner narratives they help construct. The result may be subtle or not so. Sometimes it amounts to an operatic performance. You can attend it passively, or you can intervene, modify the score to a greater or smaller extent, and produce unpredicted results.

To address the nature and composition of the orchestras within and the kinds of music they can make, I will invoke the tripartite arrangement I outlined for image making. The signals with which images are constructed originate from three sources: *the world around the organism,* from where data are collected by specific organs located in the skin and some mucosae; and two distinct components of *the world inside the organism, the old chemical/visceral compartment* and *the not so old musculoskeletal frame and its sensory portals.* It is common for accounts of mental events to privilege the world around as if nothing else were part of the mind or contributed significantly to it. It is also common for those accounts that do factor in the interior to fail to make the distinction I am making here between the evolutionarily ancient world of chemistry and viscera and the evolutionarily more recent world of the musculoskeletal frame and sensory portals.

It is often said that these "sources" are "wired" to the central nervous system and that the central nervous system makes maps and composes images out of material it receives. But this would be a misleading oversimplification of what is taking place. The relationships between the nervous system and the body are anything but simple.

First, the three sources indicated above contribute very different material to the nervous system. Second, the "wiring" from the three sources is usually seen as comparable, but it simply is not so. It is equivalent only in the sense that all three sources can generate electrochemical signals directed toward the central nervous system. In reality, however, the very anatomy and operation of the "wires" is quite distinct, especially in regard to the old chemical/visceral interior. Third, in addition to electrochemical signaling, the old interior world communicates to the central nervous system directly via even more ancient, purely chemical signals. Fourth, the central nervous system can respond *directly* to signals from the interior, especially in regard to the old interior world, thus acting on the source of the signals. In most instances, the central nervous system does not act *directly* on the outside world. *The "interior" and the nervous system form an interactive complex; the "exterior" and the nervous system do not.* Fifth, all the sources communicate with the central nervous system in a "graded" fashion so that messages are transformed as the signals are processed from their "peripheral" origins to the central nervous system. Reality is far messier than one would wish.[3]

The astonishing wealth of our mental processes hinges on images based on contributions from these worlds but assembled by different structures and processes. The exterior world contributes images that describe the perceived structure of the universe that surrounds us within the limits of our sensory devices. The old interior is the main contributor of the images we otherwise know as feelings. The new interior brings to the mind images of the overall, more or less global structure of the organism and contributes additional feelings. Accounts of mental life that fail to take these facts into consideration are likely to fall short of the mark.

To be sure, the images can be modified, added on to, and interconnected, resulting in an enrichment of mind processes. But the images that serve as a substrate for the transformations and combinations originate in three distinct worlds, and their respective and distinctive contributions need to be factored in.

Image Making

Image making of any sort, from simple to complex, is the result of the neural devices that assemble maps and that later allow maps to interact so that combined images generate ever more complex sets and come to represent the universes external to the nervous system, inside and outside the organism. The distribution of maps and corresponding images is not even. The images related to the interior world are first integrated in brain-stem nuclei, although they are re-represented and expanded in a few key regions of the cerebral cortex such as the insular cortices and the cingulate cortices. The images related to the exterior world are integrated mostly in the cerebral cortex, although the superior colliculi have an integrative role as well.

Our experience of objects and events out in the world is naturally multisensory. The organs of vision and hearing and of touch and taste and smell are engaged as appropriate for the perceptual moment. When you are hearing a musical performance in a darkened concert hall, the engagement of the senses is not the same as when you are swimming underwater and trying to see a coral reef. The dominant sensory sources differ, but they are invariably multiple and connect to multiple dedicated sensory regions of the central nervous system— the so-called "early" auditory, visual, and tactile cortices, for example.

Interestingly, another set of brain regions known as "association" cortices accomplishes the requisite integration of the images composed in the "early" ones.

The interconnection of association cortices with early cortices is responsible for the integration. As a result, the separate components that contribute to the perception of a particular moment in time may come to be experienced as a whole. One of the components of consciousness corresponds to this large-scale integration of images. The integration occurs as a result of activating varied separate regions simultaneously *and* in sequence. It is something equivalent to editing

---

FROM CONCERT HALL TO MAPMAKING ROOM

Where are maps made? It is accurate to say that the mapmaking structures are located in the central nervous system, provided it is clear that many intermediate structures in the peripheral nervous system are preparing and preassembling material for the central neural maps. In our case, the key mapmaking structures are located at three brain tiers: several nuclei of neurons in the brain stem and in the tectum (which includes the collicular nuclei); the geniculate nuclei placed higher up in the telencephalon; and, most abundantly and expansively, in numerous regions of the cerebral cortex including the entorhinal cortex and the related hippocampal system. These regions are dedicated to the processing of specific channels of sensory information. Vision, hearing, and touch arise this way, in interconnected islands of the nervous system dedicated to a particular sensory modality. Subsequently, the signals that are first segregated according to the modality are integrated. This happens at the subcortical level—in the deep layers of the superior colliculi—and in the cerebral cortex, where signals from the varied mapping regions within each sensory stream are allowed to mix and interact. They do so by means of an intricate network of hierarchical neuronal interconnections. Thanks to this integrative operation, we can, for example, see a person whose lips are moving and simultaneously hear sounds that are synchronized to the movements of the lips.

a film by selecting visual images and bits of soundtrack, ordering them as needed, but never printing the final result. The final result happens in "mind" and on the fly; it vanishes as time moves on except for the memory residue that may stay behind, in coded form. All images of the outside world are processed in nearly parallel fashion with the *affective* responses that these same images produce by acting elsewhere in the brain—in specific nuclei of the brain stem and of the cerebral cortices that are related to body state representation, such as the insular region. Which means that our brains are busy not only mapping and integrating varied external sensory sources but simultaneously mapping and integrating internal states, a process whose result is none other than feelings.

Now just pause for a second and consider the marvel that our brains accomplish as they juggle images of so many sensory kinds, of external and internal origin, and turn them into our integrated movies-in-the-brain. By comparison, film editing is a piece of cake.

Meanings, Verbal Translations, and the Making of Memories

Our perceptions and the ideas they evoke continuously generate a parallel description in terms of language. That description is also constructed with images. All the words we use, in any language, spoken, written, or appreciated by touch, as in Braille, are made of mental images. This is true of the auditory images of the sounds of letters and words and inflections and of the corresponding visual symbol/letter codings that stand for those sounds.

But minds are made of more than direct images of objects and events and of their language translations. Also present in mind are countless other images regarding any object or event that pertain and describe their constitutive properties and relationships. The collection of images typically related to an object or event amounts to the "idea" of that object or event, the "concept" of it, the meaning of it, the semantics of it. Ideas—concepts and their meanings—can be translated into the idiom of symbols and enable symbolic thought. They can also be rendered into a special class of complex symbols, the verbal idiom. Words and sentences, the latter governed by grammatical rules, execute the translation, but the translations are image based as well. All mind is made of images, from the representation of objects and events to their corresponding concepts and verbal translations. Images are the universal token of mind.[4]

The sensory integrations accomplished during perception, the ideas that their processing prompts, and the verbal translation of many aspects of these processes can be committed to memory. We construct multisensory perceptual moments in our minds, and if all goes well, we can memorize and later recall those perceptual moments and work with them in our imagination.

Later I will take up the problem of how images come to be conscious and appear to our minds as belonging so clearly and privately to each of us. We come to *know* images thanks to the complex process of consciousness, not thanks to some mysterious "homunculus." Curiously, as we shall see in chapter 9, the process of consciousness itself relies on images. But independently of their contribution to consciousness, it is apparent that once images are made and processed even at an elementary level, they can guide actions *directly* and *automatically*. They do so by depicting targets for actions and thus enabling the image-guided muscular system to reach the target more accurately. To get the gist of this advantage, just imagine that you had to defend yourself against an enemy whose presence was just signaled by smell. How would you hit the target? Where exactly? You would be missing the well-defined spatial coordinates that vision offers us directly and that sound can also help with, especially if you are a bat!

Visual images allow organisms to act on a target with precision; auditory images allow an organism to orient itself in space, even in the dark, as we can do reasonably well and bats can do so exquisitely. All that is required is that the organism be in an awake and aware state and that the content of the images be *relevant* to the life of the organism at that particular moment. In other words, from the standpoint of evolution, images must have been able to help organisms behave efficiently even when they were simply optimizers of action control and even in the absence of complex subjectivity, thoughtful analysis, and ponderation. Once image making became possible, nature could not have failed to select them.

Enriching Minds

Our complex, infinitely rich minds are, as is so often the case in the long history of life, the result of cooperative combinations of simple elements. In the case of minds, it is not a matter of cells assembled to form tissues and organs or of genes instructing amino acids to assemble myriad proteins. The basic unit for minds is the image, the image of a thing or of what a thing does, or what the thing causes you to feel; or the image of what you think of the thing; or the images of the words that translate any and all of the above.

Earlier, I mentioned that separate image streams can be integrated to produce richer accounts of external and internal realities. The integration of images related to sight, sound, and touch is a dominant mode of mind enrichment, but integration takes many forms. It can render an object from multiple sensory perspectives, and it can also string together objects and events as they interrelate in time and space and produce the sorts of meaningful sequences we call narratives. We also know the world of narrative as the world of storytelling, a world with characters and actions and props, with villains and heroes, with dreams and ideals and desires, a world in which the protagonist of the story battles his enemies and wins the heart of the girl who has been watching the events frightened but confident that her man will prevail. Life is made of an infinity of stories, simple and complex, banal and distinct, that describe all the sound and the fury and the quiet of existences and that do signify a lot.[5]

I briefly discussed the mind's secret for narrative or storytelling: hook the separate components end to end in a moving train, the train of thought, no doubt. How does the brain accomplish this? By having different sensory regions contribute the requisite part, at the right moment, so that a *time* train can be formed; by having associative structures that coordinate the timing of the components and the composition and movement of the train. Any primary sensory region can be called on to contribute, as needed; all association cortices need to participate in the timing and dispatch functions. One particular collection of association cortices that has recently been investigated in detail constitutes the so-called default mode network. This network appears to play a disproportionate role in the process of assembling narratives.[6]

Image processing also allows the brain to *abstract* images and uncover the schematic structure underlying a visual or sound image, or, for that matter, the integrated images of movements that describe a feeling state. In the course of a narrative, for example, a related visual or auditory image can be placed instead of the most predictable one, thus giving rise to a visual or auditory *metaphor*, a means to *symbolize* objects or events, in visual or auditory terms. In other words, to begin, the original images are important in and of themselves and as the grounding of our mental life. However, their manipulation can yield novel derivations.

The incessant language translation of any image that cruises in our minds is possibly the most spectacular mode of enrichment. Technically, the images that serve as vehicles for the language tracks travel in parallel to the original images being translated. They are added images, of course, translated derivations of the originals. The process is especially delightful—or maddening—for those of us with a multilingual background: we end up with multiple and parallel verbal tracks and the mix and match of words can be great fun or exasperating.

Just like the codes of cells that yielded tissues and organs, and the nucleotide codes that yielded proteins, the sounds of an alphabet that can be heard and represented in tactile or visual fashion compose the words in our minds and the words in speech and signing. Given a certain set of rules for the combination of sounds into words and for the arrangement of words according to the specific set of grammatical rules, the entire scope of our minds can be described without end.

A Note on Memory

Most everything available in our newly minted mental images is open to internal recording, whether one likes it or not. The fidelity of the recording depends on how well we attended to the images in the first place, which in turn depends on how much emotion and feeling were generated by their traversal in the stream of our mind. Many images stay on record, and substantial portions of the record can be played back, that is, recalled from the files and reconstructed, more or less accurately. Sometimes the recollection of the old material is so refined that it even competes with the new material now being generated.

Memory is present in single-celled organisms, where it results from chemical changes. There, the fundamental use of memory is the same as in complex organisms: to help recognize another living organism or situation and either approach it or avoid it. We, too, put chemical/

single-celled memory to this simple use and benefit from it. This is the kind of memory present, for example, in our immune cells. We benefit from vaccines because once we expose our immune cells to a potentially dangerous but inactivated pathogen, the cells can identify that pathogen when they next encounter it and will attack it mercilessly when it attempts to gain a foothold in our organisms.

The memories that hallmark our minds follow the same general principles except that what we memorize are not chemical modifications occurring at the molecular level but rather temporary modifications occurring in chains of neural circuits. The modifications are related to elaborate images of every sensory sort, experienced in isolation or as part of the narratives that flow in our minds. The problems that nature solved on the way to making image learning and recall possible are monumental. The solutions that nature found, at molecular, cellular, and system levels, are also admirable. At systems level, the solution most directly relevant to our discussion—the memory of images, for example, the memory of a scene that we perceive in visual and auditory terms—is achieved by converting explicit images into a "neural code" that will later allow, by working in reverse, a more or less complete reconstruction in the process of image recall. The codes represent, in non-explicit form, the actual content of images and their sequences and are stored in both cerebral hemispheres, in association cortices of the occipital, temporal, parietal, and frontal regions. These regions are interconnected, via two-way hierarchical circuits of neural cables, with the collection of "early sensory cortices" where the explicit images are first assembled. During the process of recall, we end up reconstructing a more or less faithful approximation of the original image, using reverse neural pathways, which operate from code-holding regions and produce effects within the explicit image-making regions, essentially where the images were first assembled. We have called this process retroactivation.[7]

A now famous brain structure called the hippocampus is a major partner in this process and is essential for producing the highest level of image integration. The hippocampus also allows the conversion of temporary codings into permanent ones.

Loss of the hippocampus in both cerebral hemispheres disrupts the formation of and access to long-term memory of integrated scenes. Unique events are no longer recalled even if objects and events can still be recognized outside a unique context. One is able to recognize a house as a house, but not the particular house where one has lived. The contextual, episodic knowledge acquired by personal, individual experience is no longer accessible. The generic, semantic knowledge is still recoverable. Herpes simplex encephalitis used to be a prominent cause of such a disabling loss, but Alzheimer's disease has now become the most frequent culprit. Specific cells within the hippocampal circuitry and its gateway, the entorhinal cortex, are compromised by Alzheimer's disease. The gradual disruption no longer permits effective learning or recall of integrated events. The result is a progressive loss of spatial and temporal orientation. Unique people, events, and objects can no longer be recalled or recognized. No new ones can be learned.

It is now clear that the hippocampus is an important site for neurogenesis, the process of generating new neurons that become incorporated in the local circuitry. New memory formation partly depends on neurogenesis. Interestingly, it is known that stress, which impairs memory, reduces neurogenesis.

The learning and recall of motor-related activities rely on different brain structures, namely, the cerebellar hemispheres, the basal ganglia, and the sensorimotor cortices. The critical learning and recall required for a musical performance or for the practice of sports rely on such structures in close association with the hippocampal system. Motor and non-motor image processing can be harmonized in concordance with their typical coordination in everyday activities. The images corresponding to a verbal narrative and the images corresponding to a set of related movements often occur together in real-time experience, and although their respective memories are made and held in different systems, they can be recalled in integrated fashion. Singing a song with lyrics requires the time-locked integration of varied fragments of recall: the melody that guides the singing, the memory of the words, the memories related to the motor execution.

Image recall opened new possibilities for mind and behavior. Once images were learned and recalled, they helped organisms recognize past encounters with objects and kinds of events, and by assisting reasoning, they helped organisms behave in the most precise, effective, and useful manner.

Most reasoning requires an interplay between what current images show as *now* and what recalled images show as *before*. Effective reasoning also requires the anticipation of what comes after, and the process of imagination necessary to anticipate consequences also depends on past recall. Recall helps the conscious mind with the processes of thinking, judging, and deciding—in brief, with tasks that we face on any day of our lives and on any matter of our lives from the trite to the sublime.

Recall of past images is essential for the process of imagination, which, in turn, is the playground of creativity. Recalled images are also essential for the construction of narratives, the storytelling that is so distinctive of human minds and that uses current as well as old images along with language translations of almost anything being narrated in our internal moviemaking. The meanings derived from the facts and ideas associated with the diverse objects and events included in narratives are further illuminated by the structure and course of the narrative itself.

The same story line—the same protagonists, same place, same events, same outcome—can yield different interpretations and thus have different meanings depending on the way it is told. In mental terms, the order of the introduction of objects and events and the nature of the respective descriptions relative to magnitude and qualification are decisive for the interpretation we make of the narrative, for how it will be stored in memory, and for how it will be later retrieved. We are incessant narrators of stories about almost anything in our lives, mostly about the important things but not only, and we happily color our narratives with all the biases of our past experiences and of our likes and dislikes. There is nothing fair and neutral about our narratives unless we go to the effort of reducing our preferences and prejudices, which we are well advised to do on things that matter for our lives and the lives of others.

A considerable amount of brainpower has been assigned to the search engines that both automatically and on demand can bring back remembrances of our past mental adventures. This process is critical because so much of what we commit to memory concerns not the past but the anticipated future, the future that we have only imagined for us and for our ideas. That imaginative process, which, in and of itself, is a complicated stew of current thoughts and old thoughts, of new images and old recalled images, is also unfailingly being committed to memory. The creative process is being recorded for possible and practical future use. It can come tumbling back to our present, ready to enrich our pleasure with a supplementary moment of happiness or deepen our suffering after a loss. This simple fact alone justifies the exceptional status of humans among all living beings.[8]

The constant search and sweep of our memories of past and future enable us, in effect, to intuit possible meanings of current situations and to *predict* the possible future, immediate and not so immediate, as life unfolds. It is reasonable to say that we live part of our lives in the anticipated future rather than in the present. Possibly this is one more consequence of the nature of homeostasis, with its constant projection beyond the present, in search of what comes next.

ENRICHING MINDS

Integration of images at multiple cortical sites including entorhinal cortex and the related hippocampal circuitry;

Image abstraction and metaphor;

Memory: image based learning and mechanisms of recall; search engines and prediction of the immediate future based on continuous memory searches;

Building of concepts from images of objects and events including the class of events known as feelings;

Verbal translation of objects and events;

Generation of narrative continuities;

Reasoning and imagination;

Construction of large-scale narratives integrating fictional elements and feelings;

Creativity.

7

AFFECT

The aspect of mind that dominates our existence, or so it seems, concerns the world around us, actual or recalled from memory, with its objects and events, human and not, as represented by myriad images of every sensory stripe, often translated in verbal languages and structured in narratives. And yet, a remarkable yet, there is a parallel mental world that accompanies all those images, often so subtle that it does not demand any attention for itself but occasionally so significant that it alters the course of the dominant part of the mind, sometimes arrestingly so. That is the parallel world of *affect*, a world in which we find *feelings* traveling alongside the usually more salient images of our minds. The immediate causes of feelings include (a) the background flow of life processes in our organisms, which are experienced as *spontaneous* or *homeostatic* feelings; (b) the *emotive responses* triggered by processing myriad sensory stimuli such as tastes, smells, tactile, auditory, and visual stimuli, the experience of which is one of the sources of qualia; and (c) the emotive responses resulting from engaging *drives* (such as hunger or thirst) or *motivations* (such as lust and play) or *emotions,* in the more conventional sense of the term, which are action programs activated by confrontation with numerous and sometimes complex situations; examples of emotions include joy, sadness, fear, anger, envy, jealousy, contempt, compassion, and admiration. The emotive responses described under (b) and (c) generate *provoked feelings* rather than the spontaneous variety that arises from the primary homeostatic flow. Of note, the felt experiences of emotions are unfortunately known by exactly the same name as the emotions themselves. This has helped perpetuate the false notion that emotions and feelings are one and the same phenomenon, although they are quite distinct.

Affect is thus a wide tent under which I place not only all possible feelings but also the situations and mechanisms responsible for producing them, responsible, that is, for producing the actions whose experiences become feelings.

Feelings accompany the unfolding of life in our organisms, whatever one perceives, learns, remembers, imagines, reasons, judges, decides, plans, or mentally creates. Regarding feelings as occasional visitors to the mind or as caused only by the typical emotions does not do justice to the ubiquity and functional importance of the phenomenon.

Most every image in the main procession we call mind, from the moment the item enters a mental spotlight of attention until it leaves, has a feeling by its side. Images are so desperate for affective company that even the images that constitute a prominent feeling can be accompanied by other feelings, a bit like the harmonics of a sound or the circles that form once a pebble hits the water surface. There is no *being*, in the proper sense of the term, without a spontaneous mental experience of life, a feeling of existence. The ground zero of being corresponds to a deceptively continuous and endless feeling state, a more or less intense mental choir underscoring everything else mental. I say deceptively because the apparent continuity is built piecemeal from multiple pulses of feeling derived from the ongoing image flow.

The complete absence of feelings would spell a suspension of being, but even a less radical removal of feeling would compromise human nature. Hypothetically, if you would reduce the feeling "tracks" of your mind, you would be left with desiccated chains of sensory images of the exterior world in all the familiar varieties—sights, sounds, touches, smells, tastes, more or less concrete or abstract, translated or not in some symbolic form, namely, verbal, arising from actual perception or recalled from memory. Worse, if you had been born without the feeling tracks, the rest of the images would have traveled in your mind *un*affected and *un*qualified. Once feeling would have been removed, you would have become unable to classify images as beautiful or ugly, pleasurable or painful, tasteful or vulgar, spiritual or earthy. If no feelings were available, you might still be trained, at great effort, to make aesthetic or moral classifications of objects or events. So might a robot, of course. Theoretically, you would have to rely on a deliberate analysis of perceptual characteristics and contexts and on a brute learning effort. Except that natural learning is difficult to conceive without the properties of reward and its attendant . . . feelings!

Why is the world of affect so often neglected or taken for granted when normal life is inconceivable without it? Perhaps because normal feelings are ubiquitous but often demand little attention; luckily, circumstances in which there are no major disruptions, positive or negative, tend to be the most numerous in our lives. One other reason for the neglect of feeling: affect has a bad reputation, thanks to some negative emotions whose effects are indeed disruptive or to the siren song of some seductive emotions. The conventional contrast between affect and reason comes from a narrow conception of emotions and feelings as largely negative and capable of undermining facts and reasoning. In reality, emotions and feelings come in multiple flavors, and only a few are disruptive. Most emotions and feelings are essential to power the intellectual and creative process.

It is easy to see feelings as dispensable and even dangerous phenomena rather than indispensable supporters of the life process. Whatever the cause, the neglect of affect impoverishes the description of human nature. No satisfactory account of the human cultural mind is possible without factoring in affect.

What Feelings Are

Feelings are mental experiences, and by definition they are conscious; we would not have direct knowledge of them if they were not. But feelings differ from other mental experiences on several counts. First, their *content* always refers to the body of the organism in which they emerge. Feelings portray the organism's interior—the state of internal organs and of internal operations—and as we have indicated, the conditions under which images of the interior get to be made set them apart from the images that portray the exterior world. Second, as a result of those special conditions, the portrayal of the interior—that is, the experience of feeling—is imbued with a special trait called valence. Valence translates the condition of life directly in mental terms, moment to moment. It inevitably reveals the condition as good, bad, or somewhere in between. When we experience a condition that is conducive to the continuation of life, we describe it in positive terms and call it pleasant, for example; when the condition is not conducive, we describe the experience in negative terms and talk of unpleasantness. Valence is the defining element of feeling and, by extension, of affect.

This concept of feeling applies to the basic variety of the process and to the variety that results from having multiple experiences of the same feeling. Repeated encounters with the same class of triggering situations and consequent feelings allow us to internalize the feeling process to a smaller or greater extent and make it less "bodily" resonant. As we repeatedly experience certain affective situations, we describe them in our own internal narratives, wordless or "wordy," we build concepts around them, we bring the passions down a notch or two and make them presentable to ourselves and to others. One consequence of the intellectualization of feelings is an economy of the time and energy necessary for the process. This has a physiological counterpart. Some body structures are bypassed. My notion of "as-if body loop" is one way of achieving this.[1]

The circumstances, actual or recalled from memory, that can cause feelings are infinite. By contrast, the list of elementary *contents* of feelings is restricted, confined to only one class of object: *the living organism of their owner,* by which I mean components of the body itself and their current state. But let us dig deeper in this idea, and note that the reference to the organism is dominated by one sector of the body: the old interior world of the viscera that are located in the abdomen, thorax, and thick of the skin, along with the attendant chemical processes. The contents of feelings that dominate our conscious mind correspond largely to the ongoing actions of viscera, for example, the degree of contraction or relaxation of the smooth muscles that form the walls of tubular organs such as the trachea, bronchi, and gut, as well as countless blood vessels in the skin and visceral cavities. Equally prominent among the contents is the state of the mucosae—think of your throat, dry, moist, or just plain sore, or of your esophagus or stomach when you eat too much or are famished. The typical content of our feelings is governed by the degree to which the operations of the viscera listed above are smooth and uncomplicated or else labored and erratic. To make matters more complex, all of these varied organ states are the result of the action of chemical molecules—circulating in the blood or arising in nerve terminals distributed throughout the viscera—for example, cortisol, serotonin, dopamine, endogenous opioids, oxytocin. Some of these potions and elixirs are so powerful that their results are instantaneous. Last, the degree of tension or relaxation of the voluntary muscles (which, as noted, are part of the newer interior world of the body frame) also contributes to the content of feelings. Examples include the patterns of muscular activation of the face. They are so closely associated with certain emotional states that their deployment in our faces can rapidly conjure up feelings such as joy and surprise. We do not need to look in the mirror to know that we are experiencing such states.

In sum, feelings are experiences of certain aspects of the state of life within an organism. Those experiences are not mere decoration. They accomplish something extraordinary: a moment-to-moment report on the state of life in the interior of an organism. It is tempting to translate the notion of a report into pages of an online file that can be swiped, one at a time, telling us about one part or another of the body. But digitized pages, neat, lifeless, and indifferent, are not acceptable metaphors for feelings, given the valence component we just discussed. Feelings provide important information about the state of life, but feelings are not mere "information" in the strict computational sense. Basic feelings are not abstractions. They are experiences of life based on multidimensional representations of configurations of the life process. As noted, feelings can be intellectualized. We can translate feelings into ideas and words that describe the original physiology. It is possible, and not infrequent, to *refer* to a particular feeling without necessarily experiencing that feeling or simply experiencing a paler version of the original.[2]

When one explains what a thing is, it helps to be clear about what a thing is not. So that we can be clear about what basic feelings are not, let me say that if I now decide to go down to the beach—which means that I have to take about one hundred steps down a stair before I walk on the sand—feelings are *not* primarily about the design of the movements I will make with my limbs, or about the movements of my eyes, head, and neck, all of which are also being carried out by my body, under brain control, and about whose operations my brain is also being informed. The precise notion of feeling only applies to certain aspects of the event, namely, the energy or ease with which I climb down the stairs; the eagerness with which I may do so, and the pleasure of stepping on the sand and being next to the ocean; or, for that matter, the fatigue I might feel coming back up, a while later. Feelings are primarily about the *quality of the state of life in the body's old interior,* in any situation, during repose, during a goal-directed activity, or, importantly, during the response to the thoughts one is having, whether they are caused by a perception of the outside world or by a recollection of a past event as stored in our memories.

Valence

Valence is the inherent *quality* of the experience, which we apprehend as pleasant or unpleasant, or somewhere along the range that joins those two extremes. Non-feeling representations are well designated by terms such as "sensed" and "perceived." But the representations known as feelings are *felt,* and we are *affected* by them. This is what makes the class of experiences we call feelings unique—beside the singularity of the content of feelings, that is, the body to which the brain belongs.

The deep origins of valence go back to early forms of life prior to the emergence of nervous systems and minds. But the immediate antecedents of valence are to be found in the ongoing state of life in the organism. The "pleasant" and "unpleasant" designations correspond, in a principled manner, to whether the underlying "global" state of the body is generally conducive to the continuation of life and to survival, and to how strong or weak that life trend happens to be at a given moment. Malaise signifies that something is not right with the state of life regulation. Well-being signifies that homeostasis is within the effective range. In most circumstances, there is nothing arbitrary in the relationship between the quality of the experience and the physiological state of the body. Even depression and manic states do not fully escape this rule, because basic homeostasis remains aligned to some extent, with negative or positive affect. However, pathological states such as masochism are an exception because situations of self-induced injury can be experienced as pleasurable, at least in part.

The feeling experience is a natural process of evaluating life relative to its prospects. *Valence "judges" the current efficiency of body states, and feeling announces the judgment to the body's owner.* Feelings express fluctuations in the state of life, within the standard range and outside it. Some states within the standard range are more efficient than others, and feelings express the degree of efficiency. Life within the central homeostatic range is a necessity; life upregulated to the flourishing edges is desirable. States outside the overall homeostatic range are pernicious, and some are so pernicious that they will kill you. Examples include ungainly metabolism during a generalized infection or accelerated metabolism in an overactive, manic state.

Given that we all experience feelings continuously, it is astonishing that for the most part it is so difficult to explain their nature satisfactorily. The matter of contents is about the only fairly straightforward and manageable aspect of the puzzle. We can agree on some of the events that constitute feelings, on the sequence in which they occur, and even on how events are distributed and sequenced in our bodies. In response to the big jolt of an earthquake, for example, one can sense the premature heartbeat that came fuller and earlier than normally and called attention to itself, or the dry mouth that came at the same time or just before or just after, or the tightened throat perhaps. A simple study from Riitta Hari's laboratory, in Finland, confirms the observations that several of us have long been making and agrees with the brilliant intuitions of poets. It shows that a large group of human beings consistently identified certain regions of the body as being engaged during their typical feeling experiences relative to both general homeostatic and emotional situations.[3] The head, the chest, and the abdomen were the most commonly engaged theaters of feeling. They are indeed the stages on which feelings are created. Wordsworth would have been pleased. He did write about "sensations sweet, felt in the blood, and felt along the heart," those sensations that, as he said, passed into the "purer mind, with tranquil restoration."[4]

Curiously, the precise feelings that comparable situations evoke may well be tuned by cultures. Apparently, the nervousness of students before an exam can be experienced by German students as butterflies in the stomach and by Chinese students as a headache.[5]

Kinds of Feelings

At the beginning of this chapter, I mentioned the main physiological conditions that result in feelings. The first condition produces spontaneous feelings. The other two yield provoked feelings.

Feelings of the spontaneous kind, the homeostatic feelings, arise from the background flow of life processes in our organisms, a dynamic ground state, and constitute the natural backdrop of our mental lives. They have a limited variety because they are closely tied to the humming of the living organism and to the necessarily repetitious routines of life management. Spontaneous feelings signify the overall state of life regulation of an organism as good, bad, or in between. Such feelings apprise their respective minds of the ongoing state of homeostasis, and for that reason I call them homeostatic. It is their business to "mind" homeostasis, literally. Feeling homeostatic feelings corresponds to listening to the never-ending background music of life, the continuous execution of life's score, complete with changes of pace and rhythm and key, not to mention volume. We are tuned to the workings of the interior when we experience homeostatic feelings. Nothing could be simpler or more natural.

The brain, however, is a permeable intermediary between the outside world—actual or memorized—and the body. When the body responds to brain messages that command it to engage in a certain sequence of actions—speed up respiration or heartbeats, contract this muscle group or another, secrete molecule X—the body alters varied aspects of its physical *configuration*. Subsequently, as the brain constructs representations of the altered organism geometries, we can sense the alteration and make images of it. This is the source of provoked feelings, the kinds of feelings that, unlike the homeostatic kind, result from a wide variety of "emotive" responses caused by *sensory stimuli* or by the engagement of *drives, motivations,* and *emotions* in the conventional sense.

The emotive responses triggered by the properties of sensory stimuli—colors, textures, shapes, acoustic properties—tend to produce, more often than not, a quiet perturbation of the body state. These are the qualia of philosophical tradition. On the other hand, the emotive responses triggered by the engagement of drives, motivations, and emotions often constitute major perturbations of organism function and can result in major mental upheavals.

The Emotive Response Process

A good part of the emotive process is hidden from view. The consequence of the hidden component is a change in the homeostatic state, and a possible change in ongoing spontaneous feelings as well.

When you hear a musical sound that you describe as delightful, the feeling of delight is the result of a rapid transformation of the state of your organism. We call that transformation emotive. It consists of a collection of actions that change the background homeostasis. The actions included in the emotive response include the release of specific chemical molecules in certain sites of the central nervous system or their transport, by neural pathways, to varied regions of the nervous system and of the body. Certain body sites—for instance, the endocrine glands—are brought into play and produce molecules capable of changing body functions on their own. The upshot of all this bustle is a collection of changes in the geometries of viscera—the caliber of blood vessels and tubular organs, for example, the distension of muscles, the change of respiratory and cardiac rhythms. As a result, in the case of delight, visceral operations are harmonized, by which I mean that the viscera act with no impediment or difficulty and the harmonized state of the body proper is duly signaled to the parts of the nervous system charged with making images of the old interior; metabolism is changed so that the ratio between energy demand and production is reconciled; the operation of the nervous system itself is modified so that our image production is made easier and abundant and our imagination becomes more fluid; positive images are favored over negative ones; one's mental guard is lowered even as, interestingly, our immune responses are possibly made stronger. It is the ensemble of these actions, as it becomes represented in the mind, that makes way for the pleasant feeling state that one describes as delight and encompasses a minimal amount of stress and considerable relaxation.[6] Negative emotions are associated with distinct physiological states, all of them problematic from the perspective of health and future well-being.[7]

The feelings newly provoked by emotive responses literally ride, physiologically speaking, on top of the wave of spontaneous, homeostatic responses, already traveling along in their natural flow. The process behind emotive responses is a far cry from the relative immediacy and transparency of the process behind spontaneous feelings.

Feelings may be more or less prominent in our minds. Minds engaged in a variety of analyses, imaginings, narratives, and decisions pay more or less attention to a particular object, depending on how relevant it may be at the moment. Not every item merits attention, and this is true of feelings as well.

Where Do Emotive Responses Come From?

The answer to this question is clear. Emotive responses originate in specific brain systems—sometimes in a specific region—responsible for commanding the varied components of the response: the chemical molecules that must be secreted, the visceral changes that must be accomplished, the movements of face, limbs, or whole body that are part of a particular emotion, be it fear, anger, or joy.

We know where the critical brain regions are located. Mostly they consist of groups of neurons (nuclei) in the hypothalamus, in the brain stem (where a region known as the periaqueductal gray is especially prominent), and in the basal forebrain (where the amygdala nuclei and the region of the nucleus accumbens are the lead structures). All of these regions can be activated by the processing of specific mental contents. We can envision the activation of a region as the "matching" of a certain content with the region. When the matching occurs, which is the same as saying that the region "recognizes" a certain configuration, the triggering of the emotion is initiated.[8]

Some of these regions do their jobs quite directly; others act via the cerebral cortex. Directly or indirectly, these small nuclei manage to reach into the entire organism, via the secretion of chemical molecules or the action of nerve pathways capable of initiating specific movements or releasing certain chemical modulators in a particular brain region.

This collection of subcortical brain regions is present in vertebrates and invertebrates but is especially prominent in mammals. It houses the means to respond to all manner of sensations, objects, and circumstances with drives, motivations, and emotions. Figuratively, you can see this as an "affective control panel," provided you do not imagine the emotions as immutable sets of actions triggered by a button. The nuclei do their job by increasing the probability that certain behaviors do occur and those behaviors tend to cluster together. But the result is not rigid. There are shades and variations, and only the essence of the pattern holds. Evolution has gradually built this apparatus. Most aspects of homeostasis that relate to social behavior depend on this set of subcortical structures.

The triggering of emotive responses occurs automatically and non-consciously, without the intervention of our will. We often end up learning that an emotion is happening not as the triggering situation unfolds but because the processing of the situation causes feelings; that is, it causes conscious mental experiences of the emotional event. After the feeling begins we may (or may not) realize why we are feeling a certain way.

There is very little that escapes the scrutiny of these specific brain regions. The sound of a flute, the orange tint in a sunset, the texture of fine wool, all produce positive emotive responses and the corresponding pleasant feelings. So does the picture of a summerhouse that was yours when you were growing up or the voice of the friend you miss. The sight or the aroma of a dish you especially enjoy triggers your appetite, even if you are not hungry, and a seductive photograph triggers lust. Encountering a crying child, you are motivated to hug her and protect her. Crude as it may seem, the same deeply ingrained biological drives will be engaged by the nice dog with plaintive eyes, spaced like a baby's. In brief, an endless number of stimuli will produce joy or sadness or apprehension, while certain stories or scenes will evoke compassion or awe; we emote when we listen to the warm and rich sound of a cello, independent of the melody being played, and to a high-pitched, rough sound, the felt outcome being agreeable in the former and disagreeable in the latter. Likewise, we emote positively or negatively when we see colors of certain hues, when we see certain shapes, volumes, and textures, and when we taste certain substances or smell certain odors. Some sensory images evoke weak reactions, others strong ones, in keeping with the specific stimulus and its participation in the history of a particular individual. In normal situations, numerous mental contents evoke some emotive response, strong or weak, and thus provoke some feeling, strong or weak. The "provocation" of emotive responses to countless image components or to entire narratives is one of the most central and incessant aspects of our mental lives.[9]

When the emotive stimulus is recalled from memory rather than actually present in perception, it still produces emotions, abundantly so. The presence of an image is the key, and the mechanism is the same. The recalled material engages emotive programs that yield recognizable corresponding feelings. There is a prompting stimulus, and it is still made up of images, only now the images are recalled from memory rather than constructed in live perception. Whatever the source, the images are used to produce an emotive response. The emotive response then transforms the background state of the organism, its ongoing homeostatic state, and the result is a provoked emotional feeling.

Emotional Stereotypes

Emotive responses generally conform to certain dominant patterns, but they are in no way rigid and stereotyped. The primary visceral changes, or the exact amounts of a certain molecule that are secreted during a response, vary from instance to instance. The overall pattern is recognizable, in its general arrangement, but is not an exact copy. Nor does the emotive response arise necessarily out of only one particular region of the brain, although certain brain regions are more likely to be engaged by a certain perceptual configuration than others. In other words, the idea of a "brain module" that would cause the emotive responses that lead to the feeling of delight, while another module would produce disgust, is no more correct than the idea that there is an emotive control panel with buttons for every emotion. The idea that the delight or the disgust would be a replica of each other at every new instantiation is also incorrect. On the other hand, the nature of the delight and the machinery that underlies its appearance are sufficiently comparable from instance to instance that the phenomena are easily recognizable in everyday experience and are traceable, albeit not rigidly, to certain brain systems, planted there by the grace of natural selection with the help of our genes and with more or fewer jitters from the environments of the womb and infanthood. To say that emotivity is fixed, however, would be an exaggeration. All manner of environmental factors can modify the emotive deployment as we develop. It turns out that the machinery of our affect is educable, to a certain extent, and that a good part of what we call civilization occurs through the education of that machinery in a conducive environment of home, school, and culture. In a curious way, what one calls *temperament*—the more or less harmonious manner with which we react to the shocks and jolts of life, in the day to day—is the result of that long process of education as it interacts with the basics of emotional reactivity that one is given as a result of all the biological factors at play during our development: gene endowment, varied developmental factors pre- and postnatal, luck of the draw. One thing is certain, however. The machinery of affect is responsible for generating emotive responses and, as a result, for influencing behaviors that, one could have innocently thought, would be under the sole control of the most knowledgeable and discerning components of our minds. Drives, motivations, and emotions often have something to add to or subtract from decisions one would have expected to be purely rational.

The Inherent Sociality of Drives, Motivations, and Conventional Emotions

The apparatus of drives, motivations, and emotions is concerned with the welfare of the subject in whose organism the responses inhere. But most drives, motivations, and emotions are also inherently social, at scales small and large, their field of action extending well beyond the individual. Desire and lust, caring and nurturing, attachment and love, operate in a social context. The same applies to most instances of joy and sadness, fear and panic, anger; or of compassion, admiration and awe, envy and jealousy and contempt. The powerful sociality that was an essential support of the intellect of *Homo sapiens* and was so critical in the emergence of cultures is likely to have originated in the machinery of drives, motivations, and emotions, where it evolved from simpler neural processes of simpler creatures. Even further back in time, it evolved from an army of chemical molecules, some of which were present in unicellular organisms. The point to be made here is that sociality, a collection of behavioral strategies indispensable for the creation of cultural responses, is part of the tool kit of homeostasis. *Sociality enters the human cultural mind by the hand of affect.*[10]

The behavioral and neural aspects of drives and motivations have been especially well studied by Jaak Panksepp and Kent Berridge in mammals. Anticipation and desire, which Panksepp subsumes under the label of "seeking" and Berridge prefers to call "wanting," are prominent examples. So is lust, both in its plain sex-related variety and in romantic love. The care and nurturing of progeny is another powerful drive complemented, on the side of those who are nurtured and cared for, by bonds of attachment and love, the sorts of bonds whose interruptions lead to panic and grief. Play is prominent in mammals and birds and is central to human life. Play anchors the creative imagination of children, adolescents, and adults and is a critical ingredient of the inventions that hallmark cultures.[11]

In conclusion, most images that enter our minds are entitled to an emotive response, strong or weak. The origin of the image does not matter. Any sensory process can constitute a trigger, from taste and olfaction to vision, and it does not really matter whether the image is being freshly minted in perception or recalled from the stores of memory. It does not matter if the image pertains to animate or inanimate objects, to features of objects—colors, shapes, the timbres of sounds—to actions, abstractions, or judgments on any of the above. A predictable consequence of processing many images that flow in our minds is an emotive response followed by its respective feeling. Thus provoked, emotional feelings are not quite about listening to the background music of life. Emotional feelings are about hearing occasional songs and sometimes full-regalia opera arias. The pieces are still executed by the same ensembles, in the same hall—the body—and against the same background: life. But given the triggers, the mind is now largely tuned to the world of our ongoing thoughts—rather than the world of the body—as we react to those thoughts and feel the reaction. From instance to instance, the musical execution varies because the execution of emotive responses and the experience of the respective feeling also vary, at least as much as the execution of a famous musical piece at the hands of different performers. But the score being played is still unmistakably the same. Human emotions are recognizable pieces of a standard repertoire.

A substantial portion of human glory and human tragedy depends on affect in spite of its modest, nonhuman genealogy.

Layered Feelings

The emotive responses to images even apply to the images called feelings themselves. The state of being in pain, of feeling pain, for example, can become enriched by a new layer of processing—a secondary feeling, as it were—prompted by varied thoughts with which we react to the basic situation. The depth of this layered feeling state is probably a hallmark of human minds. It is the sort of process likely to undergird what we call suffering.

Animals with complex brains similar to ours, as is the case with higher mammals, may well have layered feeling states as well. Traditionally, extreme human exceptionalism has denied feelings to animals, but the science of feeling has gradually shown the opposite. This is not to say that human feelings are not more complex and layered and elaborate than those of animals. How could they not be? But as I see it, the distinction in humans has to do with the web of associations that feeling states establish with all sorts of ideas and especially with the interpretations we can make of our present moment and of our anticipated future.

Curiously, layered feelings support the intellectualization of feelings to which I referred earlier. The wealth of objects, events, and ideas conjured up by ongoing feelings enriches the process of creating an intellectual description of the prompting situation.

Great poetry depends on layered feelings. The definitive exploration of layered feelings was the life's work of a novelist and philosopher by the name of Marcel Proust.

8

THE CONSTRUCTION OF FEELINGS

To understand the origin and construction of feelings, and to appreciate the contribution they make to the human mind, we need to set them in the panorama of homeostasis. The alignment of pleasant and unpleasant feelings with, respectively, positive and negative ranges of homeostasis is a verified fact. Homeostasis in good or even optimal ranges expresses itself as well-being and even joy, while the happiness caused by love and friendship contributes to more efficient homeostasis and promotes health. The negative examples are just as clear. The stress associated with sadness is caused by calling into action the hypothalamus and the pituitary gland and by releasing molecules whose consequence is reducing homeostasis and actually damaging countless body parts such as blood vessels and muscular structures. Interestingly, the homeostatic burden of physical disease can activate the same hypothalamic-pituitary axis and cause release of dynorphin, a molecule that induces dysphoria.

The circularity of these operations is remarkable. On the face of it, mind and brain influence the body proper just as much as the body proper can influence the brain and the mind. They are merely two aspects of the very same being.

Whether feelings correspond to positive or negative ranges of homeostasis, the varied chemical signaling involved in their processing and the accompanying visceral states have the power to alter the regular mental flow, subtly and not so subtly. Attention, learning, recall, and imagination can be disrupted and the approach to tasks and situations, trivial and not, disturbed. It is often difficult to ignore the mental perturbation caused by emotional feelings, especially in regard to the negative variety, but even the positive feelings of peaceful, harmonious existence prefer not to be ignored.

The roots for the alignment between life processes and quality of feeling can be traced to the workings of homeostasis within the common ancestors to endocrine systems, immune systems, and nervous systems. They go back in the mists of early life. The part of the nervous system responsible for surveying and responding to the interior, especially the old interior, has always worked cooperatively with the immune and endocrine systems within that same interior. Consider some current details of this alignment.

When a wound occurs, caused, for example, by an internally originated disease process or by an external cut, the usual result is an experience of pain. In the former case, the pain results from signals conveyed by old, unmyelinated C nerve fibers, and its localization can be vague; in the latter case, it uses myelinated fibers that are evolutionarily more recent and that contribute to a sharp and well-localized pain.[1] However, the feeling of pain, vague or sharp, is only a part of what actually goes on in the organism and, from an evolutionary point of view, the most recent part of it. What else goes on? What constitutes the hidden part of the process? The answer is that both immune and neural responses are engaged locally by the wound. These responses include inflammatory changes such as local vasodilation and a surge of leucocytes (white blood cells) toward the area. The leucocytes are called for to assist in combating or preventing infection and removing the debris of damaged tissue. They do the latter by engaging in phagocytosis—surrounding, incorporating, and destroying pathogens—and the former by releasing certain molecules. An evolutionarily old molecule—proenkephalin, an ancestral molecule and the first of its kind—can be cleaved, resulting in two active compounds that are released locally. One compound is an antibacterial agent; the other is an analgesic opioid that will act on a special class of opioid receptors—the δ class—located in the peripheral nerve terminals present at the site. The many signs of local disruption and reconfiguration of the state of the flesh are made locally available to the nervous system and gradually mapped, thus contributing their part to the multilayered substrate of the feeling of pain. But simultaneously, the local release and uptake of the opioid molecule helps numb the pain and reduces inflammation. Thanks to this neuro-immune cooperation, homeostasis is hard at work attempting to protect us from infection and trying to minimize the inconvenience, too.[2]

But there is more to tell. The wound provokes an emotive response that engages its own suite of actions, for example, a muscular contraction that one might describe as flinching. Such responses and the ensuing altered configuration of the organism are also mapped and thus "imaged" by the nervous system as part of the same event. Creating images for the motor reaction helps guarantee that the situation does not go unnoticed. Curiously, such motor responses appeared in evolution long before there were nervous systems. Simple organisms recoil, cower, and fight when the integrity of their body is compromised.[3]

In brief, the package of reactions to a wound that I have been describing for humans—antibacterial and analgesic chemicals, flinching and evading actions—is an ancient and well-structured response resulting from interactions of the body proper and the nervous system. Later in evolution, after organisms with nervous systems were able to map non-neural events, the components of this complex response were imageable. The mental experience we call "feeling pain" is based on this multidimensional image.[4]

The point to be made is that feeling pain is fully supported by an ensemble of older biological phenomena whose goals are transparently useful from the standpoint of homeostasis. To say that simple life-forms without nervous systems have pain is unnecessary and probably not correct. They certainly have some of the elements required to construct feelings of pain, but it is reasonable to hypothesize that for pain itself to emerge, as a mental experience, the organism needed to have a mind and that for that to pass, the organism needed a nervous system capable of mapping structures and events. In other words, I suspect that life-forms without nervous systems or minds had and have elaborate *emotive* processes, defensive and adaptive action programs, but not feelings. Once nervous systems entered the scene, the path for feelings was open. That is why even humble nervous systems probably allow some measure of feeling.[5]

It is often asked, not unreasonably, why feelings should feel like anything at all, pleasant or unpleasant, tolerably quiet or like an uncontainable storm. The reason should now be clear: when the full constellation of physiological events that constitutes feelings began to appear in evolution and provided mental experiences, it made a difference. Feelings made lives better. They prolonged and saved lives. Feelings conformed to the goals of the homeostatic imperative and helped implement them by making them *matter* mentally to their owner as, for example, the phenomenon of conditioned place aversion appears to demonstrate.[6] The presence of feelings is closely related to another development: consciousness and, more specifically, subjectivity.

The value of the knowledge provided by feelings to the organism in which they occur is the likely reason why evolution contrived to hold on to them. Feelings influence the mental process from within and are compelling because of their obligate positivity or negativity, their origin in actions that are conducive to health or death, and their ability to grip and jolt the owner of the feeling and force attention on the situation. A neutral, plain account of feelings as perceptual maps/images misses these critical ingredients: their valence and their power to capture one's attention.

This distinctive account of feelings illustrates the fact that mental experiences do not arise from plain mapping of an object or event in neural tissue. Instead, they arise from the multidimensional mapping of body-proper phenomena woven interactively with neural phenomena. Mental experiences are not "instant pictures" but processes in time, narratives of several micro events in the body proper and the brain.

It is conceivable, of course, that nature could have evolved in another way and not stumbled upon feelings. But it didn't. The fundamentals behind feelings are so integral a part of the maintenance of life that they were already in place. All that was needed in addition was the presence of mind-making nervous systems.

Where Do Feelings Come From?

To imagine how feelings appeared in evolution, it is helpful to consider what life regulation would have been like before their arrival. Simple organisms, with one cell only or many, already had an elaborate homeostatic system in charge of procuring and incorporating sources of energy, enacting chemical transformations, eliminating wastes, toxic and otherwise, replacing structural elements that were no longer functioning well and reconstructing others. When the integrity of the organism was threatened by injury, organisms could mount a multipronged defense that included the release of specific molecules and protective motions. In brief, the integrity of the organism could be maintained against all odds.

In the simplest living organisms, there was no nervous system and not even a commanding nucleus, although there were precursors to interacting organelles in the cytoplasm and a cell membrane. As noted earlier, when nervous systems finally appeared about 500 million years ago, they were "nerve nets," simple networks of neurons whose design best resembled the reticular formation of the current brain stem of vertebrates, ours included. Nerve nets were largely in charge of running the star function of the respective organisms: digestion. In lovely beasts known as hydras, nerve nets took care of locomotion—by which I mean swimming—reacted to other objects, commanded mouths to open, and executed peristalsis. Hydras were and are the ultimate floating gastronomical systems. These nerve nets were probably not capable of producing maps or images of either the outside or the inside world, and consequently the probability of their producing minds is low. Evolution would take millions of years to remedy this limitation.

Plenty of developments beneficial to homeostasis had been happening before nervous systems appeared. First, certain molecules already signified the favorable or unfavorable state of life in cells, an ability that applied all the way down the life scale to bacterial cells. Second, what is now known as the innate immune system had made its debut in early eukaryotes. All organisms with a body cavity, such as amoebas, have innate immune systems, but only vertebrates have adaptive immune systems. An adaptive immune system is a system that can be taught, trained, and boosted, for example, by vaccines.[7] Keep in mind that immune systems belong to the special class of global organism system that includes the circulatory system, the endocrine system, and the nervous system. Immunity defends us from the harm of pathogens and the ensuing damage. It is one of the earliest sentinels of organism integrity and a major contributor to valence. Circulation fulfills the homeostatic mandate by distributing energy sources and helping remove waste products. Endocrine systems adjust subsystem operations so as to fit whole-organism homeostasis. The nervous system gradually assumes the role of master coordinator for all the other global systems, while it also manages the relationships between the organism and its surrounding environment. The latter role hinges on a key development of the nervous system: the world of minds, where feelings loom large and imagination and creativity become possible.

In the scenario I currently favor, life was regulated at first without feelings of any sort. There was no mind and no consciousness. There was a set of homeostatic mechanisms blindly making the choices that would turn out to be more conducive to survival. The arrival of nervous systems, capable of mapping and image making, opened the way for simple minds to enter the scene. During the Cambrian explosion, after numerous mutations, certain creatures with nervous systems would have generated not just images of the world around them but also an imagetic counterpart to the busy process of life regulation that was going on underneath. This would have been the ground for a corresponding mental state, the thematic content of which would have been valenced in tune with the condition of life, at that moment, in that body. The *quality* of the ongoing life state would have been felt.

To begin, even if the rest of the nervous system of such creatures would be very simple, just capable of producing simple maps of varied sensory information, the introduction, in such a mix, of obligate information about the "life-favorable or life-unfavorable" state of the organism would lead to more advantageous behavioral responses than previously available. Creatures equipped with this novel element, a simple qualifier juxtaposed with the image of certain places, or objects, or other creatures, would gain an automated guide as to whether they should approach or avoid those certain places or things or creatures. Life would be better run and possibly last longer, making reproduction more likely. Then organisms equipped with the gene formulas responsible for this novel and beneficial feature would certainly win in the evolutionary selection game. The feature would inevitably spread in nature.

We have no way of knowing exactly when and how in evolution the actual emergence of feelings took place. All vertebrates have feelings, and the more I think of social insects, the more I suspect that their nervous systems generate simple minds with early versions of feeling and consciousness. A recent study favors this view.[8] One thing is for certain: the processes that *supported* feelings *after* minds emerged had been in place long before and included the mechanisms necessary to generate the hallmark component of feelings—valence.

As I see it, then, early life-forms were able to sense and respond and had the undergirding of feelings but not feelings as such, or minds, or consciousness. To arrive at what we call minds, feelings, and consciousness, evolution required a number of critical structural and functional increments that largely occurred within nervous systems.

Simpler creatures than we are, including plants, sense and respond to stimuli in their environments.[9] Simpler creatures also fight forcefully to maintain their physical integrity—but not plants, because they largely lack movement, being encased in cellulose. You can hardly punch back if you are immobile. Sensing, responding, and forceful defenses against all manner of physical threats, however, which are indispensable parts of the great and variegated story of life, are not comparable to the mental phenomena we call minds, feelings, and consciousness.

Assembling Feelings

The facts discussed so far provide a rationale for feelings and outline some critical processes behind them, namely, a scaffolding for valence. Here I point to some conditions on the side of nervous systems that probably play a complementary role in the physiology of valence.

It has become apparent that a substantial amount of the information that contributes to valence emerges in an unusual setting: a *continuity* of body structures and nervous structures. I have used other terms to explain this idea, for example, a "bonding" of body and brain, or a "compact" or "fusion" of body and brain. The term "continuity" adds another nuance.[10] In the experience of feeling, there is little or no anatomical and physiological *distance* between the object that generates the critical contents, the body, and the nervous system, which is traditionally seen as the recipient and processor of the information. The two parties, object/body and processor/brain, are surely contiguous and, in many unexpected ways, continuous. This enables them to engage in a rich interplay, and we are beginning to understand how they do so. The interplay encompasses molecular and neural operations on specific tissues and the corresponding reactions.

Feelings are *not* neural events alone. The body proper is critically involved, and that involvement includes the participation of other important and homeostatically relevant systems such as the endocrine and immune systems. Feelings are, through and through, *simultaneously and interactingly,* phenomena of *both* bodies and nervous systems.

Purely neural and purely mental phenomena would not have the ability to seize and capture the subject in the intense and forceful way that is the hallmark of strong feelings, both positive and negative. Purely mental or purely neural phenomena should not and do not fit the bill, do not provide what is needed for complex creatures to sail forth.

The Continuity of Bodies and Nervous Systems

Conventionally, chemical and visceral signals from the internal milieu use the peripheral nervous system to make their way from body to brain. Also according to convention, the central nervous system nuclei and the cerebral cortices are then responsible for the remainder of the process, that is, for actually concocting feelings. These descriptions are outdated, trapped in an early history of neuroscience that has remained untouched and incomplete for decades. A number of studies reveal several odd features that can be found in the body-brain connection and whose significance for the process of generating feelings is tantalizing. In brief, body and nervous system "communicate" using the "blendings" and "interactions" of structures that the continuity of bodies and nervous systems permits. I do not object to using the term "transmission" to describe a march of signals *within* neural pathways. But the notion of "body to brain transmission" is problematic.

If there is no distance between body and brain, if body and brain interact and form an organismic single unit, then feeling is *not* a perception of the body state in the conventional sense of the term. Here the duality of subject-object, of perceiver-perceived, breaks down. Relative to this part of the process, there is unity instead. *Feeling is the mental aspect of that unity.*

Duality does come back in, however, at a different point of the complex process of brain-body interaction. When images of the body frame and its sensory portals are formed, and when images of the spatial positions occupied by viscera are referred to that overall frame and placement within it, it becomes possible to generate a mental perspective of the organism, a set of separate images that is distinct from sensory images of the exterior (visual, auditory, tactile) *and* from the emotions and feelings they provoke. A duality sets in then, images of the "body frame and sensory-portal activity" to one side and, to the other, the rest of the images, those of the exterior *and* of the interior. That is the duality related to the process of subjectivity that I will take up in the chapter on consciousness.[11]

To date, some of the best accounts of the physiology of feelings have relied on a unique relationship between the source of what is felt— life-related activities inside the organism—and the nervous system, which, conventionally, is presumed to fabricate feeling just as it fabricates vision or thinking. But these accounts capture only part of reality and do not factor in a dramatic fact: the relationship between organism and nervous system is incestuous. The nervous system is, after all, inside the organism, but not in the same detached and clear-cut way that the reader is inside a room or that my wallet is inside my pocket. The nervous system *interacts* with varied parts of the body thanks to neural pathways, which are distributed in all body structures, and thanks, in the reverse direction, to chemical molecules, which travel in the circulating blood and can gain direct access to the nervous system at a few, fancifully named checkpoints, such as the "area postrema" and the "circumventricular organs." You can picture these regions as borderless, free traffic sites, whereas everywhere else there exists a barrier—the blood-brain barrier—that impedes the movement of most chemical molecules into the brain and vice versa.

Overall, the body is given direct, untrammeled access to the nervous system, and it is the case that the body gives free access to the nervous system, often at the very same points that communicate in the brainward direction, in a sort of tit for tat that firmly closes multiple signaling loops, from body to brain, back to body, and back to brain. In other words, as a result of the information the body offers to the brain regarding its own state, the body is being modified by return post. The range of the latter responses is fairly wide. It includes the contraction of smooth muscles in varied organs and blood vessels or the release of chemical molecules that alter the operations of viscera and metabolism. In some instances, the modification is a direct reply to what the body "told" the brain, but in other instances it is independent and spontaneous.

It should be obvious that nothing faintly comparable occurs regarding the relation, for example, between the nervous system and an object we see or hear. Objects seen or heard stand removed from the sensory apparatus that is capable of mapping their features and perceiving them, in the proper sense of the term "perception." There is no natural, spontaneous interaction of the two parties. There is distance, indeed, often great distance. It takes deliberation to interfere with an object seen or heard, and the interference is executed *outside* the duet formed by object and perceptual organ. Unfortunately, this important distinction has been systematically ignored in the relevant discussions, in cognitive science and philosophy of mind. The distinction applies less well to touch and even less well to taste and smell, the *contact* senses. Evolution has developed *telesenses* with which external objects connect to us neurally and mentally first and only reach our physiological interior via the intermediate agency of the affective filter. The older contact senses reach the physiological interior more directly.[12]

One would certainly be remiss not to point out the different manner in which the brain handles events in the interior of its organism and events that are exterior to it. One would be equally remiss not to hypothesize that this difference contributes to the construction of valence as discussed thus far. Because valence is, to begin with, a reflection of the state of goodness or badness of homeostasis within a given organism, it stands to reason that the intimacy with which body and brain go about their affairs could have a hand in translating aspects of that homeostatic state into aspects of brain function and the related, ongoing mental experience. Provided, of course, that the requisite devices for the translation exist, which, as the reader will see in a moment, is indeed the case. The intimate body-brain partnership and the physiological specifics of the intimacy contribute to the construction of valence, the main ingredient behind the seize-and-capture aspect of feelings.

The Role of the Peripheral Nervous System

Does the body really *transmit* information about its condition to the nervous system, or does the body *blend in* with the nervous system so that the latter can be continuously apprised of its status? We can conclude from what we have discussed thus far that each of these two accounts corresponds to a different age in the evolution of body-brain relationships and to different levels of neural processing. The blending-in account is the only way of describing how the old interior, using old functional arrangements, interweaves body and brain. The transmission account fits well the more modern aspects of brain anatomy and function and how they capture both the old and the not so old interior.

Conventionally, in the business of homeostasis, the body is said to transmit information regarding its doings to the central nervous system, using a variety of routes that land the relevant information in the old, so-called "emotional" parts of the brain. The typical descriptions point to certain major groups of nuclei, such as the amygdala, and to some cerebral cortices in the insular region, the anterior cingulate region, and parts of the ventromedial sector of the frontal lobe.[13] Other popular designations for this collection of structures include the "limbic brain" and the "reptilian brain." One understands how these terms made their way into the literature, but their use is not very helpful today. In humans, for example, all of these "older" structures include "modern" sectors, a bit like old houses with renovated fancy kitchens and bathrooms. Nor is the operation of these brain sectors independent but rather interactive.

A bigger problem with the traditional account is that the collection of old structures outlined above is hardly the complete story. Some parts are missing, most notably the brain-stem nuclei that are critical processors of body-related information well below the level of the cerebral cortex.[14] An important example is the parabrachial nucleus.[15] Not only do these nuclei receive information about the state of the organism, but they are also originators of the emotive responses involved in drives, motivations, and conventional emotions, a good example being the nuclei in the periaqueductal gray.[16] Perhaps most sorely lacking from the conventional account is an early and even older part, and it concerns peripheral neural structures in proximity to the body proper itself. We need to amend the account.

. . .

First, it is true that the feeling-related central nervous system structures are evolutionarily older than those that concern complex cognition. But it is equally true and much overlooked that the devices in the
"peripheral" structures, those presumed to transmit body information
to the brain, are at least as old and in some cases more so. We have paid
respects to the former and neglected the latter.

In reality, the peripheral conveyances related to the feeling process
are *not* of the sort we find transmitting signals from retina to brain in
the optic nerve, or bringing signals about fine touch from the skin to
the brain using modern and sophisticated neural fibers. For one thing,
part of the process is not even neural; that is, it does not involve regular nervous firing along chains of neurons. The process is humoral:
chemical signals traveling in the blood capillaries *bathe* certain regions
of the nervous system that are devoid of blood-brain barrier and can
thus inform those brain regions *directly* about aspects of the ongoing
homeostatic state.[17]

The blood-brain barrier, as the name suggests, protects the brain
from the influence of molecules circulating in the blood. I already
mentioned two sectors of the central nervous system that have been
well-known for lacking a blood-brain barrier. They are able to receive
chemical signals directly. Those previously known sectors are the area
postrema, located on the floor of the fourth ventricle, at brain-stem
level, and the circumventricular organs, higher up in the telencephalon, located in the margins of the lateral ventricles.[18] More recently,
it has been discovered that the *dorsal root ganglia* are also devoid of
blood-brain barrier.[19] This is especially intriguing because the dorsal
root ganglia bring together the cell bodies of neurons whose axons are
distributed widely in viscera and convey body signals to the central
nervous system.

The dorsal root ganglia are located all along the spinal column, at
the level of each vertebra, one on each side of the spine, linking the
body's periphery with the spinal cord, that is, connecting peripheral

nerve fibers to the central nervous system. This is one of the routes for conveying sensory signals from limbs and torso to the central nervous system. Information about the face is also transmitted centrally by two large but lonely ganglia: the trigeminal ganglia, one on each side of the brain stem.

This finding means that while the neurons themselves work to convey peripheral signals to the central nervous system, they do not do so alone. On the contrary, they are assisted; they are modulated *directly* by molecules circulating in the blood. The signals that, for example, help generate the pain from a wound are conveyed to precisely such dorsal root ganglia.[20] Given the arrangement I just described, the signals are thus not "purely" neural. The body has its say on the process, directly, via influential chemical molecules circulating in the blood. The same influence can be exerted higher up in the system, at the level of the brain stem and the cerebral cortices. The denuding of the blood-brain barrier is one mechanism for blending body and brain. In fact, permeability may turn out to be a fairly general feature of peripheral ganglia.[21] These facts need to be factored in the scholarship of feelings.

Other Peculiarities of the Body-Brain Relationship

It has long been known that interoceptive signals are largely conveyed to the central nervous system either by neurons whose axons are devoid of myelin, the C fibers, or by neurons whose axons are very lightly myelinated, the A delta fibers.[22] This, too, is established fact, but it has been interpreted simply as an indication of the respectable evolutionary age of interoceptive systems, and no further significance has been attributed to it. My interpretation is different. Consider the following facts.

Myelin is an important conquest of evolution. It insulates axons and allows them to conduct signals at fast speed because there is no leakage of electrical current along the axon. Our perception of the world external to our bodies—what we see, hear, and touch—is now in the well-insulated, fast, and secure hands of myelinated axons. So are the skilled and rapid movements we make out in the world, by the way, and so are the high-altitude flights of our thinking, reasoning, and creativity.[23] Myelin-dependent axon firings are modern, fast, efficient, Silicon Valley like.

How odd to discover, then, that homeostasis, the indispensable apparatus of our survival, along with feelings, the precious regulatory interface on which so much of homeostasis depends, is in the hands of the electrically leaky, slow, and ancient unmyelinated fibers. How can one explain that ever-vigilant natural selection did not get rid of these inefficient and slow propeller aircraft in favor of fast jets with "high bypass" turbines?

I can think of two reasons. Let me first present the reason that goes against my line of thinking. Myelin is created, laboriously, by wrapping non-neural cells, the glial cells also known as Schwann cells, around the axon. In brief, glia (the word means "glue") not only provides the scaffolding for neural networks but also insulates some neurons. Now, because myelin is very expensive to build, in terms of energy, the costs of fitting every axon with it might have outweighed the benefits, given that the old fibers were doing a reasonable job anyway; evolution would not have bought the product, and no further significance would be attached to the lack of myelin story.

The other reason why nature would have accepted the status quo favors my line of thinking. Unmyelinated fibers actually provide opportunities that are so indispensable to the fabrication of feelings that evolution could not afford to insulate the precious cables and jettison those opportunities.

What opportunities were created by an absence of myelin? The first has to do with the openness of unmyelinated fibers to the surrounding chemical environments. Modern myelinated fibers can only be acted on by a molecule at a few points along the axon, known as nodes of Ranvier. That is where there is a gap in the myelin insulation. But unmyelinated fibers are a different story. They are like strings that can be played anywhere along their length. This would certainly favor the functional blending of body and nervous system.

The second opportunity is more tantalizing. Because they lack insulation, unmyelinated fibers that are aligned side by side—as they are, of necessity, when they constitute a nerve—are allowed to transmit electrical impulses in a process known as ephapsis. The impulses are conducted laterally, in a direction orthogonal to the length of the fiber. Ephapsis is usually not considered in the operation of nervous systems, especially nervous systems such as ours. The attention is given, with good justification, I should add, to *synapses*, the neuron-to-neuron electrochemical signaling devices on which so much of our cognition and locomotion depends. Ephapsis is an old mechanism, a thing of the past. Often textbooks no longer refer to it. But feelings are also things of the past brought to our time because they are so useful that they are in fact indispensable. Ephapsis could alter the recruitment of axons, for example, by amplifying the responses transmitted along nerve trunks. It is intriguing to consider that the fibers in the vagus nerve, the *main* conduit of neural signaling from the entire thorax and abdomen to the brain, are almost all unmyelinated. Ephapsis may well play a role in its highly significant operations.

Non-synaptic mechanisms of transmission are a reality. They can operate not only between axons but between cell bodies and even between neurons and support cells such as glia.[24]

The Neglected Role of the Gut

That so many oddities in the body-brain relationship have not been known or have been overlooked is surprising. One of the most surprising concerns the neglect accorded to the enteric nervous system, the massive component of the nervous system that regulates the gastrointestinal tract, from the pharynx and esophagus on down. It is rarely referred to in medical teachings. When it is referred to at all, it is generally treated as a "peripheral" component of the nervous system. It has not been studied in detail until recently. It is practically absent from scientific treatments of homeostasis, feelings, and emotions, and that includes my own ventures in those areas in which the references to the enteric nervous system have been overly cautious.

In reality, the enteric nervous system is central, rather than peripheral. It is huge in structure and indispensable in function. It comprises an estimated 100–600 million neurons, a number comparable to or higher than that of the entire spinal cord. The majority of its neurons are intrinsic, in the same way that the majority of neurons in the higher brain are intrinsic; that is, they are indigenous to the structure rather than hailing from somewhere else in the organism, and they do their job within the structure rather than projecting elsewhere. Only a small fraction of neurons is extrinsic, and they project to the central nervous system largely via the famous vagus nerve. There are about 2,000 intrinsic neurons for each extrinsic one, the true mark of an independent neural structure. Accordingly, the function of the enteric nervous system is largely under its own control. The central nervous system does not tell the enteric system what to do or how to do it, but it can modulate its operations. In brief, there is a continuous cross talk between the enteric nervous system and the central nervous system, although the flow of communication is largely from the gut to the higher brain.

The enteric nervous system has recently been referred to as "the second brain." This honorable ranking is due to the system's large dimension and autonomy. There is no doubt that at this point in evolution the enteric nervous system is second only to the higher brain, in structural and functional terms. There is some evidence, however, to indicate that historically the development of enteric nervous systems might actually have antedated the development of central nervous systems.[25] There are good reasons for this, and they all have to do with homeostasis. In multicellular organisms, the digestive function is a key to the processing of energy sources. Eating, digesting, extracting the requisite compounds, and excreting are complex and indispensable operations to the life of an organism. The only equally indispensable but far more simple function than digestion is respiration. But to take oxygen from the airway and return CO_2 to the ambient air is a trifle compared with all the tasks with which the gastrointestinal tract must contend.

When one looks for the appearance of gastrointestinal tracts in evolution, one finds something resembling them in primitive creatures that belong to the Cnidaria family to which I referred earlier. As noted, cnidarians look like sacks, and they literally float for a living. Their nervous systems are nerve nets, thought to represent the oldest form of a nervous system. Nerve nets resemble the modern enteric nervous system in two ways. First, they produce peristaltic movements that facilitate the flow of food-containing water into, around, and out of the organism. Second, morphologically, they are remarkably reminiscent of an important anatomical feature of the enteric nervous system of mammals, the myenteric plexus of Auerbach. While cnidarians date to the Precambrian period, structures resembling what eventually becomes the central nervous system only appear in Platyhelminthes, in the Cambrian period. It is intriguing to think that the enteric nervous system might well have been the very *first* brain.

Given my earlier comments on myelin, we should not be surprised to discover that the neurons of the enteric nervous system are *not* myelinated. The axons are bundled together and incompletely enveloped by a bulk insulation of enteric glia. This design may well allow for ephaptic conduction, the orthogonal axonal interactions that we mentioned in relation to the nonmyelinated neurons in the peripheral nervous system. Activity in a small number of axons would recruit neighboring fibers bundled together and lead to signal amplification. Recruitment of neighboring fibers innervating contiguous territories would produce the characteristic, vaguely localized feelings that arise from gastrointestinal operations.

Several lines of evidence suggest that the gastrointestinal tract and the enteric nervous system play an important role in feeling and mood.[26] I would not be surprised if the "global" experience of grades of well-being, for example, is importantly related to enteric nervous system function. Nausea is another example. The enteric nervous system is a major tributary to the vagus nerve, the main conduit of signals from the abdominal viscera to the brain. But there are other intriguing facts germane to the argument. Digestive disorders tend to correlate with pathologies of mood, for example, and curiously, the enteric nervous system produces 95 percent of the body's serotonin, a neurotransmitter notable for its key role in disorders of affect and in their correction.[27] Perhaps the most intriguing new fact to report here is the close relationship of the bacterial world and the gut. Most bacteria live with us in happy symbiosis, occupying space everywhere in our skin and mucosae, most abundantly at places where the skin and mucosae fold. But nowhere is the number higher than in the gut, where it reaches into the billions of organisms, more individual organisms than there are individual human cells in one entire organism. How they influence the world of feeling, directly or indirectly, is an intriguing topic for twenty-first-century science.[28]

Where Are Feeling Experiences Located?

When I survey the objects that constitute my mental domains, where do I place feelings? The answer to this question is easy: I locate my feelings within the body, as represented in my mind, often with rather complete coordinates worthy of a Global Positioning System (GPS). If I cut myself peeling potatoes, I feel the cut in my finger, and the physiological mechanisms of pain tell me where exactly the cut happened: in the skin and pulp of my left index finger. The complex process responsible for the pain—as described earlier—is at first local but continues when neural signals arrive in the dorsal root ganglia assigned to the upper limbs. Here, too, the process is not exclusively neural, in the sense that molecules circulating in the blood can influence neurons directly. In turn, so-called pseudo-unipolar neurons, whose cell bodies are inside the dorsal root ganglia, convey signals to the spinal cord, where they are mixed in complex ways within the dorsal and ventral horns of the cord, at the respective level. It is only at this point, perhaps, that conventional transmission occurs and signals travel from there upward to brain-stem nuclei, thalamus, and cerebral cortex.

A standard account would state that my brain could simply register the GPS-equivalent location of the cut, in a large light panel of the sort one finds in the command and control room of a large factory or, for that matter, in the cockpit of a modern aircraft. A light goes on at location X of panel Y. It means trouble at location X, because the person in the control room has a mind to give the signal meaning. The person in charge of monitoring the panel, or the pilot, or the robotic device designed to perform the monitoring function sounds the necessary alarm and takes corrective action. But that is probably not the way our body-brain compact does things. We do locate the pain, which is useful, of course, but no less important the emotive response to pain stops us in our tracks and is *felt*. Part of our interpretation and most of our reaction depend on feeling. We react accordingly and even knowingly, if we can.

The curious thing is that our brains also have panels, such as the factory or aircraft does, located in the somatosensory regions of the cerebral cortex, which hold maps of varied aspects of our body structure: head, torso, and limbs, and their musculoskeletal frame. Yet we do not feel pain *in* the brain panel any more than the trouble in the factory is located in the panel that signals it. We feel pain at the *source,* at *the periphery,* and that is precisely where some of the builders of valence begin their hard work. This advantageous reference requires that *the brain regions most responsible for the feeling experience— some brain-stem nuclei, insular and cingulate cortices—be coactive with the brain regions responsible for the location of the peripheral processes within the global neural map of the body, for example, the somatomotor cortices.* The mind process illuminates the contents that have to do *both* with the feeling and with the point where the process originated. The two aspects do not need to be in the same neural space and clearly are not. They are traceable to separate parts of the nervous system, active in rapid sequence and yet within largely the same time unit. Moreover, the two separate parts can be functionally linked by neural connections forming a system.

Back to my potato-peeling adventure: the local details of the loss of integrity that my body suffered are responsible for a noticeable chemical, sensory, and motor disturbance that will not leave me alone until I deal with the problem in some manner. I am not allowed to ignore or forget, because the negative valence of my feeling process forcibly takes my attention from other matters. It also nearly guarantees that I will learn the details of the event quite efficiently. There is nothing distant or detached in the contents of my mental experience. I will not peel potatoes ever again.

Feelings Explained?

What can we say confidently about feelings at this point? We can say that the uniqueness of these phenomena is closely tied to the critical homeostatic role they play. The setting for the generation of feelings is radically different from that of other sensory phenomena. The relationship between nervous systems and bodies is unusual, to say the least: the former are *inside* the latter, not merely contiguous, but in some respects continuous and interactive. As shown in the previous sections, body and neural operations fuse at multiple levels, from the periphery of the nervous system all the way to the cerebral cortices and the large nuclei subjacent to them. That, and the fact that the body and the nervous system are in a ceaseless cross talk motivated by homeostatic needs, suggest that feelings are based, physiologically, on hybrid processes that are neither purely neural nor purely bodily. These are the facts and the circumstances on both sides of the equation: the mental experience we call feeling, on one side, and the body and neural processes that are circumstantially connected to it. The further exploration of the physiology behind the neural and body aspects holds the promise of further illuminating the mental side of the equation.

We have discussed feelings as mental expressions of homeostasis and as instrumental in governing life. We have also noted that due to the machinery of affect that evolution has built around feelings, and the frequent engagement of that machinery, it is not possible to talk about thinking, intelligence, and creativity in any meaningful way without factoring in feelings. Feelings play a role in our decisions and permeate our existence.

Feelings can annoy us or delight us, but that is not what they are for if we are allowed to think teleologically for a moment. Feelings are *for* life regulation, providers of information concerning basic homeostasis or the social conditions of our lives. Feelings tell us about risks, dangers, and ongoing crises that need to be averted. On the nice side of the coin, they can inform us about opportunities. They can guide us toward behaviors that will improve our overall homeostasis and, in the process, make us better human beings, more responsible for our own future and the future of others.

Events of life that make us feel well promote beneficial homeostatic states. If we love and feel loved, if what we had hoped to achieve was indeed achieved, we call ourselves happy and lucky or both, but through no specific action of ours several parameters of our general physiology move in the beneficial direction. Our immune responses become stronger, for example. The relationship between feeling and homeostasis is so tight that it cuts both ways: the disturbed states of life regulation that define diseases are felt as unpleasant. The feelings that correspond to the representation of a disease-altered body proper are unpleasant.

It is also clear that unpleasant feelings induced by external events rather than due to primarily disturbed homeostasis actually lead to states of disturbed life regulation. Continued sadness motivated by personal losses, for example, can disturb health in varied ways—reduce immune responses and diminish the alertness that can protect us from everyday harms.[29]

Both on the good and on the bad sides of the feeling coin, feelings fit the role of motives behind the development of the instruments and practices of cultures.

An Aside on Remembrances of Feelings Past

Something that especially intrigues me about memory and feeling is the way in which, at least for some of us, so many *good* moments of the past can become, in recollection, *wonderful* moments of the past, even *extraordinary* moments of the past. Good to wonderful, wonderful to extraordinary, the transformation can be magic and entertaining. The material is reclassified and regraded. There is a sweetening of the things one recalls, such that the details become more vivid and more finely etched. For example, visual and auditory images become enhanced and the associated feelings warmer, richer in tone, so delightful to experience that the very thought of interrupting the recollection becomes painful, even though the experience now past was so positive.

What can possibly account for this transformation, one should ask? I doubt age explains it (personally, I have always experienced memory this way), although it can become more pronounced with age. Does the actual frequency of good experiences rise with age so that more of them can be recalled as excellent? Unlikely. By the way, the betterment of memories, if that is how one should call the actual process, does not result from glossing over events or skipping details. On the contrary, the details of the recalled events can even be more numerous; many images of the composition can linger quite long and are allowed to produce a stronger emotive response. Perhaps that explains the betterment, after all: a careful editing of the remembrance such that certain key images are given longer screen time and are thus allowed to cause better-rounded emotions, which, in turn, translate into deeper feelings. One thing is for certain: the abundantly positive feeling that accompanies the recollection is *not* part of the material being recollected. The feeling is newly and freshly minted as a result of the strong emotive responses that the remembrances engender. In and of themselves, feelings are never memorized and thus cannot be recollected. They can be re-created, more or less faithfully, on the fly, to complete and accompany recollected facts.

It is not the case that memories of bad moments are not stored and available. It is more a matter of how much they are allowed to play in the current mind. The detail is there, and excruciatingly painful feelings can certainly be produced from it. But perhaps the not so good memories do not gain strength with time in contrast to the good memories that replay better than on past recalls. It would be a case not of suppressing details of bad memories but of lingering less over them, thus diminishing their negativity. The upshot is a highly adaptive increase in well-being.[30] The peak-end effect described by Daniel Kahneman and Amos Tversky could contribute as well. We would be prone to creating strong memories for the more rewarding aspects of a past scene and obscure the rest. Memory is imperfect.[31]

Not everyone reports this sort of affectively positive reshaping of remembrances. Some people consider that their recollections are precisely as they should be, neither better nor worse. Predictably, the pessimists among us report a worsening. But all of this is difficult to measure and judge because the courses of our lives differ in good part for reasons that have to do with our affective styles.

Why is it important to consider such phenomena? One of the reasons has to do with the anticipation of the future. What one hopes for and how one faces the life ahead depend on how the past has been lived, not only in objective, factually verifiable terms, but also in the experience or reconstruction of the objective data in one's remembrances. Recollection is at the mercy of all that makes us unique individuals. The styles of our personalities in numerous aspects have to do with typical cognitive and affective modes, the balance of individual experiences in affective terms, cultural identities, achievements, luck.

How and what we create culturally and how we react to cultural phenomena depend on the tricks of our imperfect memories as manipulated by feelings.

9

CONSCIOUSNESS

About Consciousness

In normal circumstances, when we are awake and alert, without any fuss or deliberation, the images that flow in the mind have a perspective—ours. We spontaneously recognize ourselves as the subjects of our mental experiences. The material in my mind is mine, and I automatically assume that the material in yours is yours. We each appreciate mental contents in a distinct perspective, mine or yours. If we are jointly looking at the same scene, we instantly recognize that we have different perspectives.

The term "consciousness" applies to the very natural but distinctive kind of mental state described by the above traits. That mental state allows its owner to be the private experiencer of the world around and, just as important, to experience aspects of his or her own being. For practical purposes, the universe of knowledge, current and past, that can be conjured up in a private mind only materializes to its owner when the owner's mind is in a conscious state, able to survey the contents of that mind, in his or her own subjective perspective. This perspective is so critical to the overall process of consciousness that it is tempting to simply talk about "subjectivity" and leave behind the term "consciousness" and the distractions it tends to cause. We should resist the temptation, however, because only the term "consciousness" conveys an additional and important component of conscious states: integrated experience, which consists of placing mental contents into a more or less unified multidimensional panorama. In conclusion, subjectivity and integrated experience are the critical components of consciousness.

The purpose of this chapter is to make clear why subjectivity and integrated experience are essential enablers of the cultural mind. In the absence of subjectivity, nothing matters; in the absence of some degree of integrated experience, the reflection and discernment that are required for creativity are not possible.[1]

Observing Consciousness

The conscious state of mind has several important traits. It is awake rather than asleep. It is alert and focused rather than drowsy or confused or distracted. It is oriented to time and place. The images in the mind—sounds, visual images, feelings, you name it—are properly formed, exhibited with clarity, and inspectable. They would not be if you were under the action of "psychoactive" molecules, from alcohol to psychedelic drugs. In the theater of your mind—your own Cartesian Theater, why not—the curtain is up, the actors are onstage, speaking and moving about, the lights are on and so are the sound effects, and, here comes the critical part of the setting, there is an audience, YOU. You do not *see* yourself; you simply *sense* or *feel* that in front of the theatricals onstage there sits a sort of YOU, the subject-audience for the show, inhabiting a space facing the stage's indelible fourth wall. And I am afraid even more bizarre stuff awaits because, on occasion, you may actually feel that another part of you is, well, watching YOU as you watch the show.

Some readers will be worried, at this point, that I am falling into all sorts of traps, suggesting in this torrent of metaphors that there is an actual site in the brain that could double as a theater and be a forum for mental experience. Rest assured that such is not the case. Nor do I think there is a little you or me inside the respective brains, having the experience. No homunculus, no homunculus inside the homunculus, no infinite regress of philosophical legend. The undeniable fact, however, is that it all happens *as if* there were either a theater or a gigantic Cinerama screen, and *as if* there were a me or you in the audience. It is perfectly fine to call this an illusion provided we acknowledge that there are firm biological processes behind it and that we can use them to sketch an explanation of the phenomenon. We cannot merely dismiss it as if illusions did not matter. Our organisms, specifically our nervous systems and the bodies they interact with, do not require actual theaters or spectators. They use other tricks from the body-brain partnership to produce the same results, as we will see.[2]

What else do you get to observe as the subject of your conscious mind? You might observe, for example, that your conscious mind is not a monolith. It is composed. It has parts. The parts are well integrated, so much so that some hinge on others, but they are parts nonetheless. Depending on how you make the observation, some parts may be more salient than others. The part of your conscious mind that is most salient and tends to dominate the proceedings has to do with *images* of many sensory stripes, visual, auditory, tactile, gustatory, and olfactory. Most of those images correspond to objects and events of the world around you. They are more or less integrated in sets, their respective abundance related to the activities you are engaged in at the moment. If you are listening to music, sound images may well dominate. If you are having dinner, gustatory and olfactory images will be especially prominent. Some of the images form narratives, or parts of narratives. Interspersed with the images related to ongoing perception, there may be images being reconstructed from the past, recalled on the spot because they are pertinent to the current proceedings. They are part of memories of objects, actions, or events, embedded in old narratives or stored as isolated items. Your conscious mind also includes schemas that link up images or abstractions carried out on those images. Depending on one's mental style, one can sense these schemas and abstractions more or less clearly, by which I mean, for example, that one may construct, through a glass darkly, secondary images of movements of items in space, or spatial relationships among objects.

Flowing along this super movie-in-the-brain, there are symbols, and some of them make up a verbal track that translates objects and actions into words and sentences. For most mortals, the verbal track is largely auditory and does not need to be exhaustive: not everything gets translated; our minds are not preparing subtitles for every line of dialogue or descriptions for every sight. It is a verbal track on demand that translates images hailing from the outside world but also, of necessity, images that come from the interior, as mentioned earlier.

The presence of this verbal track is one of the remaining and by now unassailable justifications for a bit of human exceptionalism. Nonhuman creatures, respectable as they are, do not translate their images into any word even when their minds do plenty of smart things that ours may or may not do.

The verbal track is co-responsible for the narrative streak of the human mind, and for most of us it may well be its main organizer. In nonverbal, quasi-filmic ways but also with words, we tell stories nonstop, to ourselves, very privately, and to others. We even ascend to new meanings, higher than those of the separate components of the story, by virtue of so much narration.

What about the other components of the conscious mind? Well, they turn out to be the images of the organism itself. One set is made of images from the old interior world, the world of chemistry and viscera, which supports feelings, the valenced images that are so distinctive in any mind. Feelings, which originate in the background homeostatic state and in so many emotive responses generated by the very images of the outside world, are major contributors to our conscious minds. They provide the qualia element that is part of traditional discussions of the problem of consciousness. Finally, there are images from the new internal world, the world of the musculoskeletal frame and its sensory portals. The images of the skeletal frame form a body phantom onto which all other images can be placed and pegged. The result of all these coordinated imaging processes is not just a great play or symphony or film. It is an epic multimedia show.

How much these components of mind dominate our mental life— that is, command attention—depends on numerous factors: age, temperament, culture, occasion, mental style, as we all tend to give more or less play to aspects of the outside world or to the world of affect.

In normal circumstances, the intensity of the subjectivity function varies and the degree of image integration varies also. When we are passionately immersed in experiencing a narrative, or even creating it anew, the subjectivity function may be extremely subtle. It is still there, readily available, present to assume its central role promptly.

When we are entirely absorbed by what is happening to the characters in a film, for example, we are not necessarily thinking about ourselves and going through the exercise of relating our enjoyment to the presence of the subject. Why would one need to allocate extra processing effort to the "I"? The stable presence of a reference "I" is enough. But notice how if, at a given moment, a word or event in the film connects to your specific past experience and provokes a reaction— a thought, an emotive response, and a specific feeling—our "subject" comes forth into saliency; we momentarily co-experience the material in the screenplay and our own presence, now made more prominent in the conscious mind. This is even more likely to occur when we have complete control of the time needed to acquire the material. That is what happens when we read a novel or even a piece of absorbing nonfiction. We can pace the acquisition and the mental translation at will, something that does not happen in a film experience unless we abandon our spectator's posture and distract ourselves from the screen. The classical film experience, as is the case with music and with reality, imposes its acquisition tempo. Turn to literature if you really want to be free.

Finally, I need to point out that the images of the interior perform double duty. On the one hand, they contribute to the multimedia show of consciousness: they can be observed as part of the consciousness spectacle. On the other hand, these images contribute to the construction of feelings and as such help with the generation of subjectivity itself, the property of consciousness that allows us to be spectators in the first place. This may appear confusing, even paradoxical at first, but it is not. The processes are nested. Feelings provide the qualia element included in subjectivity. In turn, subjectivity permits feelings to be scrutinized as specific objects in conscious experience. The apparent paradox highlights the fact that we cannot discuss the physiology of consciousness without referring to feelings and vice versa.

Subjectivity: The First and Indispensable Component of Consciousness

Let us then set aside the most salient images of the conscious mind, the ones that largely make up the contents of stories, and concentrate on the images that construct the critical enabler of consciousness: subjectivity. The reason why I am able to describe anything that goes on in my mind and say colloquially that it "is in my consciousness" is that the images that populate my mind automatically become *my* images, images that I can attend to and inspect with greater or less effort or clarity. Without my having to lift a finger, or calling for help, I *know* that the images belong to me, the owner of my mind and of the body within which that mind is being fabricated, as I write, the owner of the living organism that I inhabit.

When subjectivity disappears—when the images in mind are no longer automatically claimed by their rightful owner/subject—consciousness ceases to operate normally. If I or the reader would be prevented from holding the manifest contents of mind in a subjective perspective, those contents would float unmoored and belong to no one in particular. Who would know they existed? Consciousness would vanish, and so would the meaning of the moment. The sense of being would be suspended.

It is intriguing that a simple trick—the subjectivity trick, which we might also call the ownership trick—can turn the image-making effort of your mind into meaningful and orienting material or, by its mere absence, make the entire mind enterprise almost useless. Clearly, if we are to understand how consciousness is made, we must understand the making of subjectivity.

Subjectivity is a process, of course, not a thing, and that process relies on two critical ingredients: the building of a *perspective* for the images in mind and the accompaniment of the images by *feelings*.

1. Building a Perspective for Mental Images

When we "see," the manifest visual contents in our minds appear to us from the perspective of our vision, specifically the approximate perspective of our eyes, as set in our heads. Precisely the same happens with the auditory images in your minds. They are formed in the perspective of your own ears, not the perspective of the ears of someone else located diagonally from you or, for that matter, from the perspective of your eyes. Likewise for tactile images: they have the exact perspective of your hand, or face, or wherever else in your body comes into direct contact with what is being touched. To be sure, one smells with one's nose and tastes with one's gustatory papillae. These facts are critical to understanding subjectivity, as we will see in a moment.

One of the main contributors to the building of subjectivity is the operation of the sensory portals within which we find the organs responsible for generating images of the outside world. The early stages of any sensory perception depend on a sensory portal. The eyes and the related machinery are a prime example: the eye sockets occupy a specific and delimited region within the body, within the head, even within the face. They have specific GPS coordinates within the three-dimensional maps of our bodies, the body phantom defined by our musculoskeletal frames. The process of seeing is far more complex than projecting light patterns onto the retina. "High end" vision begins in the retinas and continues over several stages of signal transmission and processing on to the cerebral cortices dedicated to vision. But in order to see, one first needs to *look*. Looking consists of many acts, and those acts are discharged by a complicated set of devices in and around the eyes, *not* by the retinas or visual cortices. Each eye has a shutter, a diaphragm, much like those of a camera, that control the amount of light admitted to the retina. There is also a lens, again like that of a camera. It can be automatically adjusted to bring objects into focus, our very original autofocus feature. Last, the two eyes move in varied directions, in a conjugated manner, up, down, left, and right, allowing us to survey and visually capture the universe all around, not just the universe in front of us, without having to move our heads or bodies. All of these devices are continuously sensed by our somatosensory system and produce the corresponding somatosensory images. At the very same time that we construct a visual image, our brain is also imaging the slew of movements executed by these intricate devices. In the most self-referential manner possible, they inform the mind, by means of images, of what the brain and body are in the process of doing, and they "locate" those activities within the body phantom. The body phantom images are subtle, part of the spectator side of the show. They are not as vivid as those that we describe in the conscious-ness show. The brain systems that receive information concerning the movements and adjustments necessary to achieve the "looking" pro-cess are entirely different from those that receive information about visual images per se, the basis for "seeing." The "looking" machine is *not* located in the visual cortices.

Now consider the unusual situation we are identifying here: part of the process of subjectivity is made from the same kind of material with which we construct the manifest contents held in subjectivity, specifically, *images*. But while the kind of material is the same, the source is different. Rather than corresponding to the objects, actions, or events, which normally dominate consciousness, these particular images correspond to *general images of our bodies, as a whole, caught in the act of producing those other images.* This new set of images con-stitutes a partial revelation of the process of making the manifest con-tents of mind deftly and quietly inserted along those other images. The new set of images is generated within the same body that owns those manifest contents, those that are now being shown in the multi-plex stage-screen of our brains and that consciousness will let us own and appreciate. The new set of images helps describe nothing less than the owner's body in the process of acquiring the *other* images, but unless you pay close attention, you hardly notice them.

This overall strategy achieves a complex collage of (a) the funda-mental images we experience and interpret as critical to the moment we are living in our minds and (b) the images of our own organisms in the process of constructing the said images. We pay little attention to the latter, although they are essential to construct the *subject*. We save our attention for the newly minted images that describe the fun-damental contents of mind, the contents that we need to deal with if we are to continue living. This is one of the reasons why subjectivity and, more broadly, the process of consciousness have remained such a mystery. The strings of the puppetry remain conveniently hidden, as they should. None of this requires any homunculi or mysterious magic. It is so natural and simple that the best one can do is smile with respect and admire the ingenuity of the process.

What happens when the images flowing in our minds arise from memory, in recall, rather than in live perception? This same account still applies. When recalled materials are inserted in the mind contents, they are interspersed with the ongoing percepts of the moment, and the latter, fully framed and personalized, provide the "anchor" necessary for the personal perspective.

2. Feeling: The Other Ingredient of Subjectivity

The perspective generated by the musculoskeletal frame and its sensory portals is not enough to build subjectivity. Besides sensory perspective taking, the continuous availability of feelings is a critical contributor to subjectivity. The abundance of feelings generates a rich background state that one might well call feelingness.

We discussed the process of constructing feelings in the previous chapters. Here we need to consider how feelings join sensory perspective to produce subjectivity. Feelings are a natural and abundant accompaniment of the images held in the manifest component of consciousness. Their abundance derives from two sources. One source concerns the ongoing state of life whose homeostatic level results in states of well-being or malaise, of whatever grade. The ebb and flow of spontaneous homeostatic feelings provides for an ever-present background, a more or less pure sense of being of the sort that those who practice meditation aspire to experience. The other source of feeling is the processing of the multiple images that make up the processions of contents in our minds as they cause emotive responses and the respective feeling states. This latter process, as explained in chapter 7, relies on the presence of certain features in the images of any object or action or idea in our mind stream that manage to trigger an emotive response and thus produce a feeling. The numerous feelings produced in this manner join the ongoing stream of homeostatic feelings and ride on its wave. The result is that no set of images fails to be accompanied by a quota of feeling.

We conclude that subjectivity is assembled from a combination of organism perspective, relative to where in the body the images to be made conscious were generated, and the ceaseless construction of spontaneous and provoked feelings that are triggered by fundamental images and accompany them. When the images are properly placed in the perspective of the organism *and* are suitably accompanied by feelings, a *mental experience* ensues. Consciousness in the full sense of the term occurs, as we will see ahead, when such mental experiences are properly integrated in a broader canvas.

The mental experiences that constitute consciousness thus depend on the presence of mental images *and* on the process of subjectivity that makes such images ours. Subjectivity requires a perspective stance on the making of images and the pervasive feelingness that accompanies image processing, both of which come straight from the body proper. They result from the incessant tendency of nervous systems to sense and make maps of objects and events not only around the organism but also inside it.[3]

The Second Component of Consciousness: Integrating Experiences

Is the elaborate process of subjectivity, with its perspective and feeling components, enough to account for consciousness in the way we described in the first pages of this chapter? The answer is no. I wrote about the experience of attending a multimedia show where the YOU or the I was a spectator and where, on occasion, we could even attend the show of ourselves attending the show. Subjectivity, no matter how elaborate, is not enough. For that to happen, we need another component process, one that consists of integrating images and the respective subjectivities in a more or less wide canvas.

Consciousness in the full sense of the term is a particular state of mind in which mental images are imbued with subjectivity and experienced in a more or less extensive integrated display.[4]

Where are subjectivity and image integration accomplished? Is there one place in the brain, one region, or even one system *where* the related processes occur? The answer, as far as I can say, is no. As discussed in previous chapters, minds emerge in all of their complexity from the combined operations of nervous systems *and* their respective bodies, working under the baton of the homeostatic imperative, manifest in every cell, tissue, organ, and system and in their global articulation in each individual. Consciousness emerges from interactive enchainments related to life, and it goes without saying that being related to life, consciousness is also related to the universe of chemistry and physics that forms the substrate of organisms and within which our organisms exist.

There is no specific region or system of the brain that satisfies all the requirements of consciousness, the perspective and feeling components of subjectivity, and the integrating of experiences. Not surprisingly, the attempts to find one brain locus for consciousness have not been successful.[5] On the other hand, it is possible to identify several brain regions and systems that are unequivocally related to producing key ingredients of the process as outlined earlier: perspectival stance, feeling, and experience integration. These regions and systems participate in the process as an ensemble, entering and departing the assembly line in orderly fashion. Once again, those brain regions are not doing it alone; they work in intense cooperation with the body proper.

My hypothesis, then, is that the contributing ingredients are produced regionally and incorporated in sequential, parallel, or even superposed processions. In a typical scenario, subjectivity for a scene dominated by visual and auditory parts would require activity at multiple sites of the visual and auditory systems, in both brain-stem structures and cerebral cortices. The related evocation of images from memory would be interspersed with the main image cohort of the scene. Activity related to the feelings caused by the flowing images would be provided by nuclei in the upper brain stem, hypothalamus, amygdala, basal forebrain, and insular and cingulate cortices, in inter-action with varied sections of the body proper. As for the activity related to sensory portals/musculoskeletal frame, it would be produced in the brain stem's tectum—the superior and inferior colliculi—and in somatosensory cortices and frontal eye fields. Last, part of the coordi-nation of all these activities would take place in medial cortical regions, especially in posteromedial cortices, assisted by thalamic nuclei.

The process related to the integration of experiences requires the narrative-like ordering of images and the coordination of those images with the subjectivity process. This is achieved by association cortices of both cerebral hemispheres arranged in large-scale networks, of which the default mode network is the best-known example. Large-scale networks manage to interconnect noncontiguous brain regions by fairly long bidirectional pathways.

In brief, varied parts of the brain, working in close interaction with the body proper, make images, generate feelings for those images, and co-reference them to the perspective map, thus accomplishing the two ingredients of subjectivity. Other parts of the brain command a sequen-tial highlight of images. Each highlight occurs *at* its sensory source, thus contributing to a broad display of images that moves along in time but not in place. The images do not need to be moved about in the brain. They come into subjectivity and integration by virtue of local, sequential highlight. Smaller or higher numbers of images and narra-tives can be processed at each time unit, and that determines the scope of integration at each moment. The separate brain regions, and many of the body regions that assist them, are interconnected by actual neural pathways and can be traced to neuroanatomical structures and systems. Still, the panoramic integrated experience with which I began this chapter—the theatrical or film exhibition observed by a subject (you, me)—is to be found not in one single brain structure but rather in more or less numerous time series of frames being activated piecemeal, not unlike the multiple frames that make up a physical movie. But note that when I used the metaphor of movie-in-the-brain earlier, I was simply considering the making and ordering of plain images in a narrative. I was not considering the even more complex process of imbuing them with subjectivity and amplifying the scope of the integration to a vaster multidimensional canvas where space is dependent on time.

The picture that emerges in this hypothesis is one in which the upper tier of the process is, through and through, dependent on local neural systems, on pathways that interconnect them, and on interactions with the body. The overall process unfolds in time but results from exquisite contributions that are firmly rooted in specific, localizable organism operations. The process is inconceivable without the contributions of the organism's periphery via direct chemical action on the peripheral nervous system and central neural structures. It requires a host of brain-stem nuclei and other telencephalic nuclei. It requires cerebral cortices of all evolutionary ages, old and new. To privilege one of these neural sectors over the others in the making of consciousness would be folly, and so would ignoring the presence of the body proper, which the nervous system is in charge of serving.[6]

From Sensing to Consciousness

There is merit to the idea that consciousness, in the broad sense of the term, is widely available in numerous living species. The issue, of course, is the "kind" and amount of consciousness exhibited by other species. There is not much doubt that bacteria and protozoa sense and respond to the conditions of their environment. So do paramecia. Plants respond to temperature, hydration, and amount of sunlight by slowly growing roots or turning their leaves or flowers. All of these creatures continually *sense* the presence of other living creatures or of the environment. But I resist calling them conscious, in the traditional meaning of the word, because that traditional meaning is tied to the notions of mind and feeling, and in turn I have linked mind and feeling to the presence of nervous systems.[7] The creatures mentioned above do not have nervous systems, and nothing suggests that they have mental states. In short, a mental state, a mind, is a basic condition for conscious experiences to exist in the traditional meaning. When that mind acquires a point of view—a subjective point of view, that is—consciousness proper may begin.

So much for beginnings. Where consciousness ends, as we have seen, is very high up, in the stratosphere of complex, integrated, multisensory experiences to which subjectivity is applied. These experiences refer to both the ongoing world external to the subject and complex worlds of yore, namely, the world of the subject's past experience, as assembled from recalled memories. They also refer to the world of the subject's current body state, which, as I pointed out earlier, is the anchor for the subjectivity process and thus a crucial element of consciousness writ large.

The fact that there is a long physiological and evolutionary distance between the sensing and irritability of plants and single cells, on the one hand, and mental states and consciousness, on the other, does not signify that sensing and mental states and consciousness are unrelated. On the contrary, mental states and consciousness depend on the elaboration, within creatures equipped with nervous systems, of strategies and mechanisms present in simpler, pre-neural creatures. This begins to happen, evolutionarily, in nerve bundles, in ganglia, and in nuclei within central nervous systems. Eventually, it happens in brains in the proper sense.

Between the phenomena of cellular sensing, as a basic level of this natural process, and mental states in the full accession of the term sits an intermediate critical level, made up of the most fundamental of mental states: feelings. Feelings are core mental states, perhaps *the* core mental states, those that correspond to a specific, foundational content: *the internal state of the body within which consciousness inheres*. And because they pertain to the varied quality of the state of life within a body, feelings are necessarily *valenced;* that is, they are good or bad, positive or negative, appetitive or aversive, pleasurable or painful, agreeable or disagreeable.

When feelings, which describe the inner state of life *now,* are "placed" or even "located" *within the current perspective of the whole organism,* subjectivity emerges. And from there on, the events that surround us, the events in which we participate, and the memories we recall are given a novel possibility: they can actually *matter* to us; they can affect the course of our lives. Human cultural invention requires this step, that events be made to matter, that they be automatically classified as beneficial or not to the individual to which they belong. Conscious, owned feelings permit a first diagnosis of human situations as problematic or not. They animate the imagination and excite the reasoning process on the basis of which a situation will be found to be problematic or a false alarm. Subjectivity is required to drive the creative intelligence that constructs cultural manifestations.

Subjectivity was able to endow images, minds, and feelings with novel properties: a sense of ownership related to the particular organism in which these phenomena were happening; the *mineness* that allows entry into the universe of individuality. Mental experiences gave minds a new impact, an advantage to countless living species. And with humans, mental experiences were direct levers in the deliberate construction of cultures: the mental experiences of pain, suffering, and pleasure became foundations for human wants, stepping-stones of human inventions, in sharp contrast to the collection of behaviors assembled up to that point by the workings of natural selection and genetic transmission. The gulf between the two sets of processes—biological evolution and cultural evolution—is so large that it makes one overlook the fact that homeostasis is the guiding power behind both.

Images cannot be *experienced,* in and of themselves, until they are part of a *context* that includes *specific sets of organism-related images,* those that naturally tell the story of how the organism is being perturbed by the engagement of its sensory devices with a particular object. Where the object is—out in the world, elsewhere in the body proper, or recalled from a memory created by a prior imaging of something internal or external to the organism—is not important. *Subjectivity is a relentlessly constructed narrative.* The narrative arises from the circumstances of organisms with certain brain specifications as they interact with the world around, the world of their past memories, and the world of their interior.[8]

The essence of the mysteries behind consciousness is made of this.

An Aside on the Hard Problem of Consciousness

The philosopher David Chalmers focused the investigation of consciousness when he identified two problems in consciousness studies.[9] In practice, both problems relate to the understanding of how the organic material of nervous systems could give rise to consciousness. The first problem, termed "easy," referred to the complex but decipherable mechanisms that allow brains to construct images and the instruments with which images can be manipulated such as memory, language, reasoning, and decision making. Chalmers thought that ingenuity and time would solve the easy problem. I believe he was right. Wisely, in my view, he did not have any issue with mapmaking and image making.

The "hard" problem identified by Chalmers was to understand why and how the "easy" parts of our mental activity became conscious. In his words, "Why is the performance of these mental functions [the functions described under the easy problem] accompanied by experience?" So the hard problem refers to the issue of mental experience and to how mental experience can be constructed. When I am conscious of a certain percept—for example, you the reader, in front of me, or the image of a painting with its form and color and suggested depth—I know automatically that either image is mine, that it belongs to me and to no one else. As noted earlier, this aspect of the mental experience is known as subjectivity, but the mere mention of subjectivity does not conjure up the functional ingredients with which I have just proposed to build it. I am referring to the *quality of the mental experience,* feelingness, and to the *placement of feelingness within the perspectival frame of the organism.*

Chalmers also wants to know why experience is "accompanied" by feelings. Why does the feeling that accompanies sensory information exist at all?

In the explanation that I propose, experience is itself partly generated from feelings, and so it is not really a matter of accompaniment. Feelings are the result of operations necessary for homeostasis in organisms such as ours. They are integrally present, made from the same cloth as other aspects of mind. The homeostatic imperative that pervaded the organization of early organisms led to the selection of programs of chemical pathways and specific actions that ensured the maintenance of organism integrity. Once there were organisms with nervous systems and image-making ability, brain and body cooperated to image those complex multistep programs of integrity maintenance in a multidimensional manner, and that gave rise to feelings. As mental translators of the homeostatic advantages of the chemical and action programs, or lack thereof, relative to varied objects, their components, and situations, feelings let the mind know about the current state of homeostasis and thus added another layer of valuable regulatory options. Feelings were a decisive advantage that nature would not have failed to select and use as consistent accompaniment to mental processes. The answer to Chalmers's question is that *mental states naturally feel like something because it is advantageous for organisms to have mental states qualified by feelings.* Only then can mental states assist the organism in producing the most homeostatically compatible behaviors. In fact, complex organisms such as ours would not survive in the absence of feelings. Natural selection made certain that feelings would become a permanent feature of mental states. For more detail on how life and nervous systems produced feeling states, the reader can refer to the previous chapters and recall that feelings arose from a series of gradual, body-related processes, bottom up, from simpler chemical and action phenomena accumulated and maintained over evolution.

Feelings changed the evolution of carbon-based creatures such as we are. But the full impact of feeling only took place later in evolution, when feeling experiences were inserted and appreciated in the broader perspective of a subject and made to matter to the individual. Only then did they begin to influence imagination, reasoning, and creative intelligence. This occurred only when the otherwise isolated experience of feeling was located within the imagetically constructed subject.

The hard problem is about the fact that if minds emerge from organic tissue, it may be hard or impossible to explain how mental experiences, in effect, *felt* mental states, are produced. Here I suggest that the interweaving of perspectival stance and feelings provides a plausible explanation for how mental experiences arise.

PART III

THE CULTURAL MIND AT WORK

10

ON CULTURES

The Human Cultural Mind in Action

All mental faculties intervene in the human cultural process, but in the last five chapters I chose to highlight the ability to make images, affect, and consciousness, because cultural minds are not conceivable without such faculties. Memory, language, imagination, and reasoning are leading participants in cultural processes, but require image making. As for the creative intelligence responsible for the actual practices and artifacts of cultures, it cannot operate without affect and consciousness. Curiously, affect and consciousness also happen to be the faculties that got away, forgotten in the throes of the rationalist and cognitive revolutions. They deserve special attention.

By the end of the nineteenth century, the role of biology in the shaping of cultural events was acknowledged by Charles Darwin, William James, Sigmund Freud, and Émile Durkheim, among others.[1] At about the same time, and into the early decades of the new century, biological facts were invoked by a number of theorists (among them Herbert Spencer and Thomas Malthus) to defend the application of Darwinian thinking to society. These efforts, generally known as social

Darwinism, resulted in eugenic recommendations in Europe and in the United States. Later, during the Third Reich, biological facts were misinterpreted and applied to human societies with the goal of producing a radical sociocultural transformation. The result was a horrifying and massive extermination of specific human groups targeted because of their ethnic background or political and behavioral identity. Biology was unfairly but understandably blamed for this inhumane perversion. It would take decades for the relation between biology and cultures to become an acceptable subject of scholarship.[2]

By the last quarter of the twentieth century and thereafter, sociobiology and the discipline it spawned, evolutionary psychology, have made a case not only for a biological perspective on the cultural mind but for the biological transmission of culture-related traits.[3] The latter efforts concentrated on the relationship between cultures and the genetic replication process. The fact that the worlds of feeling and reason are in endless interplay and that cultural ideas, objects, and practices are inevitably caught in their accommodations and contradictions has not been the focus of those efforts (although evolutionary psychologists have included the action component of the world of affect—such as emotions—in their proposals). The same applies to the topic I privilege in this book: the ways in which the cultural mind copes with human drama and exploits human possibilities, and the manner in which cultural selection completes the cultural mind's job and complements the achievements of genetic transmission. I am not favoring affect and human drama, to the exclusion of other participants in the cultural process. I am simply focusing attention on affect—and feeling in particular—in the hope that it can be more clearly incorporated in accounts of the biology of cultures. To achieve this, I must insist on the role of homeostasis and of its conscious deputy—feeling—in the cultural process. In spite of all the historical forays of biology into the world of cultures, the notion of homeostasis, even in the conventional and narrow sense of life regulation, is absent from classical treatments of culture. As noted earlier, Talcott Parsons did mention homeostasis when he considered cultures from the perspective of systems, but in his account homeostasis was unrelated to feelings or to individuals.[4]

How does one connect the state of homeostasis to the making of a cultural instrument capable of correcting a homeostatic deficit? As I suggested, the bridge is provided by feeling, a mental expression of the homeostatic state. Because feelings mentally represent a currently salient state of homeostasis and because of the upheaval they can generate, feelings operate as motives for engaging the creative intellect, the latter being the link in the chain that is responsible for the actual construction of the cultural practice or instrument.

Homeostasis and the Biological Roots of Cultures

In the first chapters of this book, I wrote that several important aspects of human cultural responses were foreshadowed in the behaviors of living organisms that are simpler than we are. The astonishingly effective social behaviors of those organisms, however, were not invented by formidable intellects or motivated by feelings resembling ours. They resulted from the natural and extraordinary way in which the life process copes with the homeostatic imperative, the blind champion of advantageous individual and social behaviors. The formulation I propose to address the biological roots of the human cultural mind specifies that homeostasis has been responsible for the emergence of behavioral strategies and devices capable of ensuring life maintenance and flourishing, in simple as well as complex organisms, humans included. In early organisms, homeostasis generated the *precursors* to feeling and subjective perspective in the absence of mental processes. Neither feelings nor subjectivity was present, only the mechanisms that were necessary and sufficient to help regulate life prior to the development of nervous systems and minds.

All of these mechanisms relied on naturally selected chemical molecules—within the precursors to the endocrine and immune systems—and naturally selected action programs. Many of these mechanisms have been well conserved to this day and we know them as emotive behaviors.

In later organisms, after nervous systems emerged, minds became possible and, within them, feelings along with all the images that represented the exterior world and its relation to the organism. Such images were supported by subjectivity, memory, reasoning, and eventually verbal language and creative intelligence. The instruments and practices that constitute cultures and civilizations in the traditional sense emerged thereafter.

Homeostasis achieved the survival and flourishing of the individual and helped create the conditions for it to persist and reproduce.[5] At first, living organisms dealt with such goals without recourse to nervous systems and minds, but later they used minded and deliberative ways. The most expedient strategies among the available plethora were selected in evolution and, as a result, were genetically maintained over generations. In simpler organisms, the selection was made from options naturally generated by processes of autonomous self-organization; in complex organisms, the selection came to be cultural; it was made from options produced by subjectively directed invention. The level of complexity varied, but the unstated basic homeostatic goals remained the same—survival, flourishing, and potential reproduction. This is a good reason why practices and instruments that in one way or another exhibit "sociocultural" characteristics arose early and more than once in evolution.

In unicellular organisms, such as bacteria, we find that rich social behaviors, without any deliberation by the organism, reflect an implicit judgment of the behavior of others as conducive or not to the survival of the group or of individuals. They behave "as if" they judge. This is early "culture" achieved without a "cultural mind." Here is an early manifestation of the sort of schematic solution that wisdom and clear reason would come to use and prescribe once full-fledged minds could think through a problem whose essence was comparable.

In social insects, multicellular creatures with elaborate nervous systems, the complexity of the "cultural" behaviors is higher. The behavioral practices are more complex, and there is also a production of concrete instruments, for example, the architectural colony as a physical entity. Numerous other species produce artifacts as well— elaborate nests, simple tools. The important distinction, of course, is that the nonhuman cultural manifestations are often the result of well-established programs deployed in appropriate circumstances and in largely stereotyped fashion. The programs have been assembled over eons, by natural selection, under the control of homeostasis, and have been transmitted by genes. In the case of un-brained, un-nucleated bacteria, the command centers for the deployment of programs are located in the cell's cytoplasm; in the case of multicellular metazoan species such as insects, the command centers are located in the nervous system, where they have been shaped by the genome.

As one contemplates evolution and its branches, one can glean border transitions between pre-mental and post-mental organisms. To some extent, those borders correspond to the distinction between "pre-cultural" behaviors and "truly cultural" behaviors and minds. There is an intriguing alignment of purely genetic evolution with the former and mixed but largely cultural evolution with the latter.

Distinctive Human Cultures

The picture we can draw for the human cultural mind and its cultures differs on numerous counts. The governing imperative is still the same—homeostasis—but there are more steps on the way to achieving results. First, capitalizing on the established existence of a corpus of simple social responses in existence since bacterial life

began—competition, cooperation, simple emotivity, collective production of instruments of defense such as biofilms—the many species
in the lineage that preceded us evolved and genetically transmitted
a class of *intermediate mechanisms* capable of producing complex,
pro-homeostatic emotive responses that are also, more often than
not, social responses. The critical component of those mechanisms is
lodged in the machinery of affect described in chapter 7. It is responsible for deploying drives and motivations and responding to varied
stimuli and scenarios emotively.

Second, capitalizing on the fact that intermediate mechanisms
produce complex emotive responses and their subsequent mental
experiences—feelings—homeostasis could now act transparently.
Feelings became motives for new forms of response, engendered by
the uniquely rich creative intellect and motor ability of humans. These
new forms of response were able to control physiological parameters
and achieve the sorts of positive energy balances that are essential for
homeostasis. But the new forms of response were innovative in yet
another way. The ideas, practices, and instruments of human cultures
could be transmitted culturally and were open to cultural selection. In
addition to the genetic antecedents that allowed organisms to respond
in a particular way under certain circumstances, cultural products now
marched in part to their own drum, surviving or becoming extinct on
their merits, as guided by homeostasis and the values it determined.
This innovation leads us to a third, and no less important feature of
the relation between feelings and culture: *feelings could also act as
arbiters of the process.*

Feelings as Arbiters and Negotiators

The natural process of life regulation orients living organisms so that
they operate within the range of parameters compatible with life main-

tenance and flourishing. The heroic process of maintaining life requires a precise, herculean process of regulation, in individual cells as well as in whole organisms. In complex organisms, feelings play a critical role in that process at two levels. First, as we have seen, when organisms are forced to operate outside the well-being range and they drift into disease and toward death. As that happens, feelings are powerful disturbances that inject into the thinking process a striving for a desirable homeostatic range. Second, besides generating concern and compelling thinking and action, feelings serve as arbiters of the quality of response. Ultimately, feelings are the judges of the cultural creative process. This is because, in good part, the merits of the cultural inventions end up being classified as effective or not so by a feeling interface. When feeling pain motivates a solution to make the pain vanish, the reduction of the pain is indicated by a feeling—of pain subsiding. That is the critical signal to decide on whether the effort worked. Feelings and reason are involved in an inseparable, looping, reflective embrace. The embrace can favor one of the partners, feeling or reason, but it involves both.

In sum, the categories of cultural response that are part of today's repertoire would have succeeded at correcting dysregulated homeostasis and returning organisms to prior homeostatic ranges. It is reasonable to think that those categories of cultural response survive because they accomplished a useful functional goal and were accordingly selected in cultural evolution. Curiously, the useful functional goal would also have increased the power of certain individuals and, by extension, of groups of individuals relative to others. Technologies are a good example of this possibility: think of navigation expertise, trading skills and accounting, printing, and now digital media. To be sure, the added power is an advantage for those who control it. But the achievement of power is fueled by properly felt ambition and is followed by a rewarding affect. The idea that cultural instruments and practices were conceived with the purpose of managing affect—

and, by extension, producing homeostatic corrections—is plausible. It also goes without saying that the cultural selection of successful instruments and practices can have repercussions on the frequencies of genes.

Assessing the Merits of an Idea

How does this idea of the workings of the cultural mind fit the actual manifestations of human cultures? The case for varied early technologies, no doubt some of the first cultural manifestations, is easy to make. Toolmaking—for hunting, defense, and attack—shelter, and clothing are good examples of how intelligent inventions responded to fundamental needs. Those needs first came to be known to the respective humans by way of spontaneous homeostatic feelings such as hunger, thirst, extreme cold or heat, malaise, and pain, which pertain to the management of *individual* life states and signify deficient homeostasis. The need for food—and the search for food sources such as meat, that would yield energy reasonably fast, the need for shelter to provide protection from intemperate climate and create a safe haven for infants and children; the need to defend self and group from predators and foes—all were efficiently signaled by feelings related, for example, to parent-infant bonding and attachment and to fear. These feelings were then acted upon by knowledge, reason, and imagination, in brief, by creative intelligence. In the same vein, disease states, from wounds and fractures to infections, were detected primarily by homeostatic feelings and dealt with by new technologies that became gradually more efficient and that history came to know as medicine.

Most provoked feelings result from engaging emotions that relate not just to the isolated individual but to the *individual in the context of others*. Situations of loss result in sadness and despair, whose presence solicits empathy and compassion, which stimulate the creative imagination to produce counters to the sadness and despair. The result can be simple—a set of caring gestures, the protection provided by physical contact—or complex: a song or a poem. The ensuing resumption of homeostatic conditions opens the way for recruiting more complex feeling states—gratitude and hope, for example—and a subsequent reasoned elaboration over those feeling states. There is a close association between beneficial forms of sociality and positive affect, and an equally close association of both with a suite of chemical molecules in charge of regulating stress and inflammation such as endogenous opioids.

It is not possible to imagine the origin of the responses that became medicine or any of the principal artistic manifestations outside an affective context. The sick patient, the abandoned lover, the wounded warrior, and the troubadour in love were able to *feel*. Their situations and their feelings motivated intelligent responses, in themselves and in other participants in their respective situations. Beneficial sociality is rewarding and improves homeostasis, while aggressive sociality does the opposite. But it should be clear that I am not confining the arts to a therapeutic role today. The pleasures that can be derived from an art piece are still related to their therapeutic origin but can soar into new intellectual regions where they are joined by complexities of ideas and meanings. Nor am I suggesting that all cultural responses are intelligent and well-organized accomplishments that necessarily produce an effective answer to the original plight.

Other examples of emotive reaction and cultural response include, on the positive side of the ledger, yearning to alleviate the suffering of others and taking pleasure in discovering a means to do so; delighting in finding ways to improve the lives of others ranging from the offer of material goods to playful inventions that result in happiness; taking pleasure in the consideration of nature's mysteries and attempting to solve them. This is how many cultural ideas, instruments, practices, and institutions were probably born, modestly and in small groups.

Over time they became places of worship, books of wisdom, exemplary novels, institutions of learning, declarations of principle, and charters of nations.

On the negative side, violence toward and from other human beings played an inordinate role. Its leading cause was the engagement of a neural apparatus of emotions whose development possibly came to a peak in great apes and whose shadow continues to loom over the human condition.

Such violence came largely from males, and it did not have to be justified by hunger or group-related fights for territory. It could target females and the young as well as other adult males. Humans inherited the potential for these modes of behavior, which were highly adaptive for a long stretch of human history, and biological evolution has not succeeded in eradicating the potential for violence.[6] Cultural evolution, thanks to human creativity, has actually expanded the range of expressions of violence. The Florentine tradition of *calico storico* as well as rugby and football are good examples. Physical violence remains present in some competitive sports, as heirs to Roman gladiator spectacles, and is consistently rehearsed by varied forms of entertainment in movies, television, and the Internet. Physical violence is also abundantly present in the surgical strikes of modern warfare, terrorist and otherwise. As for nonphysical, psychological violence, it is present via unrestrained abuses of power well exemplified by the invasion of privacy made possible by modern technologies.

One of the jobs of cultures has been to tame the beast that has been so often present and that remains alive as a reminder of our origins. Samuel von Pufendorf's definition of culture addresses these points: "the means by which human beings overcome their original barbarism, and through artifice, become fully human."[7] Pufendorf does not mention homeostasis, but my take on his words is that barbarism leads to suffering and disturbed homeostasis, while cultures and civilizations aim at reducing suffering and thus restore homeostasis by resetting and constraining the course of the affected organisms.

Today a significant number of cultural instruments and practices turn out to be responses to grievances and violations of rights that manifest themselves not merely as factual descriptions of certain predicaments and circumstances but as powerful and energizing emotions such as anger and revolt and as the consequent feeling states. Here we find affect and reason as two components of social movements. The anthems and poetry that celebrate the crushing of enemies in bloody victories are part of the history behind that process.

From Religious Beliefs and Morality to Political Governance

Early medicine was not prepared to address the traumas of the human soul. But a case can be made that religious beliefs, moral systems and justice, and political governance were largely aimed at those traumas and at recovering from their consequences. I see the development of religious beliefs as most closely related to the grief of personal losses, which forced humans to confront the inevitability of death and the myriad ways in which it could come about: accidents, diseases, violence perpetrated by others, and natural catastrophes, anything but old age, a rare condition in prehistoric times. But many traumas of the human soul were inflicted by public events in the social space, and religious beliefs were appropriate responses in varied ways.[8]

The response to losses and to the grief caused by violence was varied, and depending on the subject, it included empathy and compassion but also rage and more violence. We can comprehend that the grief would have been countered by an adaptive conception of suprahuman powers in the form of gods capable of resolving large-scale conflicts and putting an end to a high degree of violence. In an animistic period of cultures, such gods would have been asked to help not just with personal suffering but with the protection of personal and community property—crops, domesticated animals, vital territory. Later, in the case of monotheistic cultures, the belief in such entities would eventually take the form of one single God capable, for example, of explaining losses in justifiable and even acceptable terms. Ultimately, the promise of a continuation of life beyond death might entirely void the negative effects of any losses and offer another meaning for them.

Nowhere is the feeling and homeostatic motivation of religious beliefs and practices more clearly spelled out than in Buddhism. The founder of Buddhism, the perceptive, well-informed, and philosophically genial prince Gautama, identifies suffering as a corrosive aspect of human nature and sets out to eliminate it by reducing its most frequent cause: the desire to indulge pleasures by whatever means and the inability to achieve such pleasures consistently. Gautama proposes salvation—freedom from the homeostatic insecurity of striving for perpetuity while realizing the futility of the effort—by sidestepping self entirely in exchange for the very experience of being.

Cold reason would also use sentinel feelings to prompt its contribution. The repeated encounter with instances of suffering caused by stealing, lying, betrayals, and erratic discipline would have been a powerful prompt for the invention of codes of conduct whose recommendations and practice would result in reductions of suffering.

I see the development of moral codes, justice systems, and political governance, beginning with the egalitarian arrangements of early human tribes and continuing with the complicated administration formulas of the Bronze Age kingdoms or the Greek and Roman Empires, as closely related to the development of religious beliefs in connection to feelings and, through feelings, to homeostasis. Gods, and eventually one God, are a means of transcending the erratic interests of humans and of seeking a *dis*interested authority that can be impartial, trusted, and respected. Of note, over the past two decades, the investigation of neural and cognitive phenomena related to morality and religion has made contact with feelings and emotions, as shown in work from our research group and in the work of Jonathan Haidt, Joshua Greene, and Lianne Young. These findings are especially well discussed by Mark Johnson and by Martha Nussbaum from the point of view of moral philosophy.[9]

Another important homeostatic route for the development of religious practices refers to situations of large-scale threat and disaster. Examples include the confrontation with major climate calamities— floods and droughts—earthquakes, plagues, and wars.[10] They would engage social motivations and would result in powerful and cooperative collective behaviors. Fear, dread, and anger would be immediate results and compromise homeostasis, but cooperative group support would follow along with attempts to comprehend, justify, and respond to the situation constructively. Some responses would include behaviors later incorporated in religious, artistic, and governance practices. Wars constitute a special case because they can prompt both constructive remedies and endless cycles of violence begetting violence. There is nothing to be added to what Homer, the Mahabharata, and Shakespeare's history plays illustrate on this issue.

Whether homeostasis is approached from the solace and consolation angle or from the benefits produced by collective organization and sociability, religion and homeostasis can be persuasively linked in terms of their origins and historical endurance, the latter being indicative of robust cultural selection. I suspect that Émile Durkheim—who placed the roots of religion in collective rituals of tribal peoples rather than in the assuaging of individual or small group sufferings—might agree. Such collective behaviors, as Durkheim commented, unleashed powerful, rewarding emotions and feelings. The collective behaviors of Durkheim's tribal peoples, however, are likely to have been prompted by homeostatic instabilities in the first place. The homeostatically stabilizing outcome for the individuals in the group would still apply.

Karl Marx is supposed to have talked about religion as "the opium of the masses" (although he did not quite say that; he said, instead, that religion was "the opium of the people," the "masses" probably being a post-Leninist retrofit). What could be more homeostatically inspired than the notion of prescribing opioids to treat human pain and suffering?

Marx also wrote, in advance of that famous sentence, "Religion is the sigh of the oppressed creature, the heart of a heartless world, and the soul of soulless conditions." Here is an interesting mixture of social analysis and probing scrutiny of the cultural mind. It combines his rejection of religion with the pragmatic recognition that religion can be a soulful refuge in a dehumanized and soulless world. Noteworthy, considering that Marx had no idea of how dehumanized and soulless the world would become, especially the world he was responsible for inspiring. Noteworthy most of all because of the transparent linkage of life state, feelings, and cultural responses.[11]

The fact that the history of religions is rife with episodes in which religious beliefs led and still lead to suffering, violence, and wars, hardly humanly desirable outcomes, in no way contradicts the homeostatic value that such beliefs did have and clearly still have for a large part of humanity.

Finally, just as in the case of artistic endeavors, I need to make clear that I do not see religions as mere therapeutic responses. That the initial motivation of religious beliefs and practices was related to homeostatic compensation is both plausible and likely. How such early attempts evolved is another matter. The intellectual constructions that followed have gone beyond the goal of consolation to serve as instruments of inquiry and formulation of meaning where the compensation element is only a vestige. Practical goals were followed by philosophical explorations of the meaning of humans and universe.

The Arts, Philosophical Inquiry, and the Sciences

The arts, philosophical inquiry, and the sciences make use of an especially broad range of feelings and homeostatic states. How can we imagine the birth of the arts and not picture the reasoning of one individual working on the resolution of a problem posed by a feeling—the artist's own or someone else's? That is how I conceive of the development of music and dance, painting, and eventually poetry, the theater, and cinema. All of these art forms were also tied to intense sociality because motivating feelings often came from the group and the effect of the arts transcended the individual. Beyond the satisfaction of individual affective needs in the original participants, the arts played important roles in the structure and coherence of groups, in multiple settings, from religious ceremonies to the preparation for war.

Music is a powerful inducer of feelings, and humans gravitate toward certain instrumental sounds, modes, keys, and compositions that produce rewarding affective states.[12] Music making provided feelings for multiple occasions and purposes, feelings that could effectively cancel suffering and offer consolation, personal and of others. The feelings generated by music were probably also used for seduction and for pure playful and personal contentment. Humans certainly built flutes, with five holes, no less, by as early as about fifty thousand years ago. Why would they have bothered to do so if they had not found a rewarding use for the effort? Why would they have engaged in the time-consuming effort of perfecting these newly fashioned tools, rejecting some and accepting others after testing their effects? In those early days of music making, they would have been discovering that certain kinds of sounds—instrumental and vocal—produced predictably agreeable or disagreeable effects. In other words, the emotive response caused by a wind sound—vocal or fluted—and the ensuing feeling would have been a welcome discovery of soothing or seductive effects; the rough, raspy sound of sticks and stones rubbed together would not. Moreover, as sounds were added together, they could prolong the pleasantness and produce other layers of effect, for example, mimic objects and events in an appropriate sequence and begin to tell a story.

The specific emotivity tied to sounds is comparable to the emotivity found for colors, shapes, or surface textures. The physical nature of such stimuli constitutes an emblematic signal of the goodness or badness of the *whole* objects that typically exhibit such physical components. Those objects were consistently associated in evolution with positive or negative life states—dangers and threats or well-being and opportunities, in brief, the states that underlie pleasure or pain. We humans, along with the creatures from which we descend biologically, inhabit a universe in which objects and events, animate as well as inanimate, are not affectively neutral. On the contrary, as a consequence of its structure and action, any object or event is naturally *favorable* or *unfavorable* to the life of the individual experiencer. Objects and events influence homeostasis positively or negatively and, as a result, yield positive or negative feelings. Just as naturally, the separate *features* of objects and events—their sounds, shapes, colors, textures, motions, time structure, and so forth—become *associated, by learning,* with the positive or negative emotions/feelings linked to the whole object/event. This is, I believe, how the acoustic features of certain sounds come to be described as "pleasant" or "unpleasant." The characteristics of a sound, which are a part of an object/event, acquire the affective significance that the *whole* event had for the individual. That systematic bond, between the isolated feature and the affective valence, lives on, independently of the original association that gave rise to it. This is why we end up saying that the sound of a cello is beautiful and warm: the acoustic characteristics of the particular sound were once part of the experience of pleasantness caused by an entirely different object. The high-pitched sound from a trumpet or violin may be experienced as disagreeable or frightening for the same sort of reason. We draw on long-established associations—many of which preceded the appearance of humans and are now part of our standard neural equipment—in order to classify musical sounds in affective terms. Humans were able to explore such associations as they constructed sound narratives and laid down all sorts of rules for the combination of sounds. [13]

By the time humans were making flutes, they had probably been putting to good use the very first musical instrument—the human voice—and perhaps the second instrument ever: the human chest, a natural cavity suitable for drumming. As for the third instrument, it was probably an actual, manufactured hollow drum.

Whether consoling or seducing, in activities that tended to involve two individuals or a group gathering for a communal event—a birth, a death, the arrival of food, the celebration of an idea, religious or otherwise, playing merrily, or marching off to tribal wars—music contributed its multipronged homeostatic effects, early and most probably often, beginning with layer upon layer of feelings and ending in ideas.[14] Music's universality and remarkable endurance seem to come from this uncanny ability to blend with every mood and circumstance, anywhere on earth, in love and in war, involving single individuals, small groups, or large groups suddenly made cohesive by the power of music. Music serves all masters as quietly as an old-world butler or as loudly as a heavy metal band.

Dance was closely linked to music, and its movements accomplished expressions of comparable sentiments—compassion, desire, the exulting joys of seduction accomplished, of love, of aggression, and of war.

The case for the homeostatic function of the visual arts—which begin with cave paintings—and for the tradition of oral storytelling in poetry, theater, and political exhortation, is not difficult to make.

These manifestations often referred to the management of life—food sources and the hunt, for example, the organization of the group, wars, alliances, loves, betrayals, envies, jealousies, and, quite often, the violent resolution of the problems faced by the participants. Paintings, and far later texts, provided signposts and pauses for reflection, warning, play, and enjoyment. They provided attempts at clarifications for what must have been confusing confrontations with reality. They helped sort out and organize knowledge. They provided meaning.

Philosophical inquiry and science developed from the same homeostatic cloth. The questions that philosophy and science aimed at answering were prompted by a large range of feelings. Suffering was prominent, no doubt, but so was the perturbation and worry caused by chronic puzzlement over the enigmas of reality—once again, the vagaries and irregularities of climate, floods, and earthquakes, the movement of the stars, the life cycles that could be observed in plants and animals and in other humans, and the odd combination of benevolent and destructive behaviors that describes the actions of so many humans personally. The destructive feelings, whose result has so often been war, have played a major role in science and technology. Repeatedly in history, war efforts have been made possible or collapsed by the success or failure of the technology and sciences that permitted the development of weapons.

There were other feelings, too, not least the pleasant feelings that resulted from the very process of attempting to solve the enigmas of the cosmos and the anticipation of the rewards that their solution would bring. Precisely the same sorts of problems and the same kind of homeostatic need would lead different humans at different times and places to formulate religious or scientific explanations for their plight. The ultimate purpose was assuaging the pain, reducing the need. The form and efficiency of the response is another issue.

The homeostatic benefits of philosophical inquiry and scientific observation are endless: in medicine, obviously, and in physics and chemistry as enablers of the technologies on which our world has long depended. They include the harnessing of fire, the invention of the wheel, the invention of writing, and the subsequent advent of written records external to the brain. The same applies to later innovations that are responsible for modernity, from the Renaissance onward, and, all along, to the ideas that have informed, for better and worse, the governance of empires and countries, as expressed, for example, in the Reformation, the Counter-Reformation, the Enlightenment, and more generally modernity.

While the lion's share of cultural achievement must go to the intelligent invention of specific solutions to varied predicaments, we must note that even the automated attempt at homeostatic correction—which is mediated by the machinery of affect—can, in and of itself, produce beneficial physiological consequences. By breaking down isolation and bringing individuals together, the simple drive to socialize generates opportunities to improve or stabilize individual homeostasis. The mechanisms of mutual grooming in mammals are an example of an instinctual pre-cultural arrangement whose homeostatic effects are significant. In strictly affective terms, grooming delivers pleasurable feelings; on the health side, it reduces stress and prevents tick infestations and the resulting diseases.

Along the exact same lines and using the same highly conserved neural and chemical mechanisms, the fellowship engendered by collective cultural manifestations induces responses that reduce stress, generate pleasure, promote increased cognitive fluidity, and more generally have beneficial effects for health.[15]

Contradicting an Idea

We can attempt to challenge my general hypothesis by addressing situations that contradict the idea and deciding whether the contradictions are real or apparent. How can we, for example, think of religious beliefs as homeostatic when religion can cause so much suffering itself? And what about cultural practices that result in self-mutilation or exorbitant weight gain?[16]

The issue of religious belief is important to consider. The positive homeostatic effect of religious belief can be documented individually—it does reduce or eliminate suffering and despair, and it does give way to varied degrees of well-being and hope. This is physiologically verifiable.[17] It is also documented that large sectors of the world population hold varied religious beliefs and the overall number of believers is actually stable or growing rather than diminishing, an indication of strong cultural selection. The hypothesis addresses not the features, or internal structure, or external consequences of beliefs, but simply the fact that individual or group losses and the attending homeostatic disruption caused by suffering can be reduced by cultural responses that involve religious beliefs. The fact that religious beliefs can *also* provoke suffering does not contradict the hypothesis. Moreover, religious beliefs generate other remarkable benefits such as social group membership whose positive homeostatic consequences are obvious. The same can be said for the music, architecture, and art that are directly attributable to religious belief and the related religious organizations. Feelings, acting in their role as arbiters, would have contributed to the persistence of ideas that promoted so many homeostatically advantageous outcomes. Cultural selection ensured the adoption of the related ideas and institutions.

Certain cultural instruments can actually worsen homeostatic regulation or even be the primary cause of dysregulation. One obvious example comes from the adoption of systems of political and economic governance that were originally meant to respond constructively to extensive social suffering but ended up producing human catastrophes. Communism, for example, accomplished precisely that. The homeostatic goal of the invention is undeniable and conforms to the hypothesis I have advanced. The results, immediately and in the long run, were something else, producing in some cases greater poverty and violent death than the world wars that flanked the dissemination of these systems. This is a paradoxical case in which rejection of injustice, a process theoretically favorable to homeostasis, leads unintentionally to more injustice and homeostatic decline. But nothing in the general hypothesis speaks to the guaranteed success of homeostatic inspiration. Success depends on how well conceived the cultural response is in the first place, on the circumstances to which it applies, and on the features of the actual implementation.

The hypothesis does specify that the success of the response is monitored by the same system that is responsible for its motivation: that is, feeling. It can be argued that the misery and suffering produced by such social systems were the cause of their demise. But why, then, did it take so long for the demise to occur? At first glance, the adoption or rejection of cultural responses depends on cultural selection. Ideally, the results of cultural responses are monitored by feelings, weighed by the collective, and judged as beneficial or harmful by a negotiation between reason and feeling. But truly beneficial cultural selection assumes certain conditions that in practice can fail. For example, in the case of governance and moral systems, it assumes democratic freedoms so that the adoption or rejection of a response is not coerced. It also assumes some sort of level playing field in terms of knowledge, reasoning, and discernment. In the cases of varied communist and fascist regimes, cultural selection had to bide its time and still does.

Taking Stock

We can venture that what we now consider true cultures quietly began in simple, single-celled life, under the guise of efficient social behavior guided by the imperative of homeostasis. Cultures only became fully worthy of the name billions of years later in complex human organisms animated by cultural minds, that is, probing and creative minds, still operating under the same powerful homeostatic imperative. In between the early, un-minded foreshadowings and the late flourishings of cultural minds stands a series of developments that can also be seen, in retrospect, as consonant with the requirements of homeostasis.

First, the mind had to be capable of representing, in the form of images, two distinct sets of data: the world exterior to the individual organism, where the *others* that are part of the social fabric loom prominently and interactively; *and* the state of the individual organism's interior, which is experienced as feelings. This capability draws on an innovation of central nervous systems: the possibility of making, within their neural circuitries, maps of objects and events that are located outside the neural circuitries. Such maps capture "resemblances" of those objects and events.

Second, the individual mind had to create a mental perspective for the whole organism relative to those two sets of representations—the representations of the organism's interior and of the world around it. This perspective is made up of images of the organism during the acts of perceiving itself and its surround, in reference to the organism's overall frame. This is a critical ingredient of subjectivity that I regard as the decisive component of consciousness. The fabrication of cultures, which requires social, collective intentions, is inconceivable without the presence of multiple individual subjectivities working, to begin with, for their own advantage—their own interests—and eventually, as the circle of interests enlarges, promoting the good of a group.

Third, once mind had begun but before it could become the cultural mind we can recognize today, it was necessary to enrich it by adding impressive new features. Among them were a powerful, image-based memory function capable of learning, recalling, and interrelating unique facts and events; an expansion of the imagination, reasoning, and symbolic thought capabilities such that nonverbal narratives could be generated; and the ability to translate nonverbal images and symbols into coded languages. The latter opened the way for a decisive tool in the construction of cultures: a parallel line of verbal narratives. Alphabets and grammars were the "genetic" tools of this latter and enabling development. The eventual invention of writing was the crowning entry into the toolbox of creative intelligence, an intelligence capable of being moved by feeling to respond to homeostatic challenges and possibilities.

Fourth, a critical instrument of the cultural mind resides with a largely unsung function: *play*, the desire to engage in seemingly useless operations that includes the moving about of actual pieces of the world, real or in toy form; the moving of our own bodies in that world, as in dancing or playing an instrument; the moving of images in the mind, real or invented. Imagination is a close partner of this endeavor, of course, but imagination does not fully capture the spontaneity, the range and reach of PLAY, to use the capitalized form that Jaak Panksepp prefers when he talks about this function. Think of play when you think about what can be done with the infinity of sounds, colors, shapes, or with pieces in Erector or Legos sets or computer games; think of play when you think of the infinitely possible combinations of word meanings and sounds; think of play as you plan an experiment or ponder different designs for whatever it is that you are planning to do.

Fifth, the ability, especially developed in humans, to work *cooperatively* with others to achieve a discernible, shared goal. Cooperativity relies on another well-developed human ability: joint attention, a phenomenon to which Michael Tomasello has devoted pioneering studies.[18] Play and cooperation are, in and of themselves, independently of the results of the respective activities, homeostatically favorable activities. They reward the "players/cooperators" with a slew of pleasurable feelings.

Sixth, cultural responses begin in mental representations but come into being by the grace of movement. Movement is deeply embedded in the cultural process. It is from emotion-related movements happening in the interior of our organisms that we construct the feelings that motivate cultural interventions. Cultural interventions often arise from emotion-related movements—of the hands, quite prominently, of the vocal apparatus, of the facial musculature (a critical enabler of communication), or of the whole body.

Last, the march from life's beginnings to the doors of human cultural development and cultural transmission was only possible due to another homeostasis-driven development: the genetic machinery that standardized the regulation of life inside cells and permitted the transmission of life to new generations.

The rise of human cultures should be credited to both conscious feeling and creative intelligence. Negative and positive feelings needed to be present in early humans, or the upper tier of the cultural enterprise such as the arts, religious belief and philosophical inquiry, moral systems and justice, science, would have lacked a prime mover. Unless the process behind what became pain was *experienced*, it would have been a mere body state, a pattern of operations in the clockwork of our organisms. The same would apply to well-being or joy or fear or sadness. To be experienced, the patterns of operations related to pain or pleasure had to be turned into feeling, which is the same as saying that they had to acquire a *mental* face, which is the same as saying that the mental face had to be owned by the organism in which it occurred, thereby becoming *subjective*, in brief, *conscious*.

Non-experienceable pain and pleasure mechanisms, by which I mean *nonconscious and nonsubjective* pain-related and pleasure-related mechanisms, clearly assisted early life regulation in an automatic and undeliberated way. But in the absence of subjectivity, the organism in which such mechanisms occurred would not have been able to consider either the mechanism or the results. The respective body states would not have been *examinable*.

The collection of questions, explanations, consolations, adjustments, discoveries, and inventions that make up the noblest part of human history required a motive. Felt pain and suffering, on their own, but especially when contrasted with felt pleasure and flourishing, did move the mind and call for action. Provided, of course, that there was something to be moved in the mind, and there certainly was, especially as *Homo sapiens* developed, in the form of the expanded cognitive and language abilities discussed earlier. In the most practical terms, that movable something was the ability to *think* beyond what could be immediately perceived and the ability to *interpret* and *diagnose* a situation, understanding causes and effects. How correct the interpretations and diagnoses were, over the ages, is not the point. Obviously, they were often incorrect. The point was having an interpretation, correct or not, firmly motivated by a strong feeling, positive or negative. On that basis, it was possible for intensely social humans to motivate the invention, individually and in the collective space, of previously nonexisting responses. This movable, mental something involves not just what we sense as reality here and now, but reality as it might have happened or as it might have been forecast to happen. I am referring to *recalled* reality, a reality that can be altered by our imagination, processed in chains of remembered images of every sensory stripe—sight, sound, touch, smell, taste—images that can be cut in pieces and moved about, playfully recombined to form new arrangements and address specific goals: the construction of a tool, a practice, an explanation. None of this is incompatible with the earlier appearance, prior to *Homo sapiens*, of some limited cultural manifestations such as stone tools.[19]

The movable something identified the relationships between certain objects, people, events, or ideas and the onset of either suffering or joy; it provided an awareness of the immediate and not so immediate antecedents to pain and pleasure; and it identified possible and even likely causes. The scale of the events could actually be quite large and have equally large consequences. History provides instances of such antecedents, for example, the social upheavals that preceded the development of major systems of religious belief such as Judaism, Buddhism, and Confucianism—for example, the disruptive wars and the terrorism of the "Sea Peoples" that brought down the civilizations of the Mediterranean in the twelfth century B.C. in a setting that probably includes devastating earthquakes, droughts, and economic and political collapse. But thousands of years before the development of the golden Axial Age cultures—the period that spans the six centuries before the Christian era and that includes the explosion of Athenian philosophy and theater—humans had been inventing all manner of social creations as a response to their feelings. The feelings were not restricted to those of loss, pain, suffering, or anticipated pleasure. Included were also responses to yearnings for social community, as an extension to larger groups of feelings that began in care of progeny, attachment, and nuclear families, as well as the drive toward objects, people, and situations capable of yielding admiration, awe, and a sense of sublimity.

The inventions prompted by feelings included music, dance, and the visual arts, along with rituals, magical practices, and busy, multitasking gods and goddesses with which humans tried to explain and resolve some puzzles of everyday life. Humans also formalized schemes of complex social organization, beginning with fairly simple tribal arrangements and progressing to the culturally structured life in the fabled kingdoms of the Bronze Age in Egypt, Mesopotamia, and China.

The mental movable something that yielded complex cultural developments also included the startling realization that on occasion no antecedent to pain or pleasure could be identified, no explanation could be found at all, there simply being pain or pleasure without any reason for either being apparent, just mystery. The resulting powerlessness, and even despair, are also likely to have been a sustained driving force behind human endeavors and have had a hand in arriving at and developing notions such as transcendence. In spite of the extraordinary triumphs of science, so much mystery remains that those forces are still durably at play in most world cultures.

Feelings focused intelligence on certain goals, increased the reach of intelligence, and refined it in such a way that it resulted in a human cultural mind. To some degree, for better and worse, feelings and the intellect they mobilized have freed humans from the absolute tyranny of genes but only to keep us under the despotic rule of homeostasis.

A Hard Day's Night

We are all familiar with the magic of the evening, with sunsets that turn into twilights, then give way to the night and the stars and the moon. We humans gather together at those bewitching hours, we talk and drink, play with children and dogs, discuss the good and bad events of the day now ending, argue about the problems of family or friends or politics, plan the next day. We still do this in any season, winter included, by a fire, real or gaslit, a likely carryover from a past long gone, for that is how the complex cultural activities of the early evening might well have started, around a simple campfire, out in the open, a starlit sky above.

The harnessing of fire dates to no more than one million years ago, probably less, and according to Robin Dunbar and John Gowlett campfires have been a practice for several hundred thousand years, possibly before *Homo sapiens* came onto the scene.[20] What was so important about the control of fire? An amazing collection of developments, as it turns out, cooking getting the star billing. Fire made way for the invention of cooking and the possibility of eating digestible and highly nourishing meats rapidly, as opposed to slowly munching on veggies for hours at a time, with little to show for it in terms of acquired energy. Bodies and their brains could now grow apace with plenty of vital proteins and animal fat to help sharpen minds in charge of the myriad tasks needed to support all this gourmet consumption. Fire-cooked food favored a specific place for eating, reduced the time needed for chewing food, and, by so doing, freed time for other activities. And this is where we discover a hidden payoff of fire: a specific setting conducive to newly minted activities. A whole tribe could gather around a campfire, not just for cooking and eating, but for socializing. Until then, the coming of darkness normally caused the brain to trigger the secretion of the hormone melatonin and the resulting ushering in of sleep. But firelight delayed melatonin secretion and increased the usable hours of the day. No one would be hunting or gathering in the early evening, and later, once agriculture began, no one would be tilling the land. The duration of a day was now extended. The day's work was done, but the community was still up and awake, ready to relax and repair in the full old sense of the word. It is not difficult to imagine conversations about troubles and successes, about friendships and enmities, about work relationships or amorous ones, no matter how simple the conversations would be, and there is no reason to assume they would be that simple once *Homo sapiens* came into its own. What better time to mend ties broken during the day or cement new connections built during the day? What better time to discipline unruly children and instruct them? And think of the open sky and its stars and how they begged for answers about what it all meant—crepuscules, flickering lights, Milky Ways, a moon that moved about in the sky and changed its shape capriciously but predictably, eventual dawns. Chanting and dancing are not difficult to imagine either, or witchcraft.

Polly Wiessner has written persuasively about firelight gatherings based on her contemporary studies with Ju/'hoansi Bushmen in southern Africa.[21] She has suggested that once foraging daytime duties were over, firelight opened the way for a productive use of the early night hours: conversations, abundant storytelling, gossip of course, the mending of what was humanly broken during a hard day's work, the cementing of social roles in small groups of humans.

The next time you enjoy sitting by a fire, ask yourself, why would humans still wish to build something as old-fashioned and often useless as a fireplace in their modern homes? The answer perhaps is that the hearth can still work in the rich cultural way it once did, that the idea of the potentially advantageous setting still produces an appropriately encouraging feeling of anticipation. Just call it magic.

11

MEDICINE, IMMORTALITY, AND ALGORITHMS

Modern Medicine

It is not difficult to glean the homeostatic relevance of most human cultural practices, but nowhere is the relevance more apparent than in medicine. From its formal beginning, thousands of years ago, the entire practice of medicine has been an exercise in the repair of diseased processes, organs, and systems, occasionally tied to magic and religion and eventually to science and technology.

The current panorama of developments in medically related science and technology is wide, and the aims range from the conventional to the delusional. At the conventional end, one finds treatments for reasonably understood diseases capitalizing on pharmacological or surgical tools made possible by recent scientific and technical progress. The history of infectious diseases is a good example. The once fatal ravages of infections have been controlled by the development of antibiotics or vaccines or both. The battle is never ending because new infectious agents appear on the scene or because old ones change so much—often as a result of antibiotic therapy—that they come to behave as badly as if they were new. The saga of new corrections never ends, however. Nature is properly defensive and evasive, but medical science does not lack ingenuity or persistence. For example, when the cause of disease is a dangerous virus that is normally carried by a certain insect species, it is now possible to change the insect's genome such that its carrier status is blocked. This is bold and new and freshly possible due to the discovery of a technique, CRISPR-Cas9, that permits successful modifications within a genome.[1] Nothing guarantees, of course, that the thwarted viruses will not mutate in response to the genetic dissuader and defy the new barrier they face by increasing their malignancy. And so it goes. Homeostasis knows how to play games of cat and mouse, and sometimes so do we.

Using the same novel techniques, we will be able to produce modifications of the human genome aimed at eliminating certain hereditary diseases. This is another laudable and potentially valuable endeavor, but there is nothing easy about it because most hereditary diseases that plague humanity are caused not by only one troublesome gene but by several, sometimes many. Genes often operate in bundled fashion, a bit like toxic mortgages. Guaranteeing that the result of an intervention does not produce dangerous and unwanted effects is easier said than done.

Far more problematic are some medically unconventional developments, for example, inducing genetic modifications aimed at guaranteeing favorable intellectual and physical traits or retarding and eliminating death. Here, too, the target of the intervention is the human germ line, and the interventions are also enabled by the bold new technique I mentioned earlier.

There are serious issues to be considered in the implementation of the latter projects. At a practical level, there are important risks involved in the manipulation of genetic material that, to date, do not appear to have been properly addressed. More fundamentally, tinkering with the natural process of evolution has unforeseen consequences for the future of humanity, in strict biological terms and in sociocultural, political, and economic terms. If the aim is eliminating a disease that produces suffering and is not associated with any benefit, there is ample justification to proceed. Medicine's classical injunction is "first do no harm," and provided the injunction is carefully observed, one should applaud the tinkering. But what if there is no disease to begin with? On what grounds is it justifiable to try to improve one's memory capacity or intellectual caliber by genetic means rather than by practicing intellectual puzzles? And what about physical traits—eye color, skin color, facial design, height? And what about the manipulation of gender ratios?

It can be argued that these are "cosmetic" changes and that cosmetic surgery has been practiced for decades with little harm and plenty of satisfied customers. (Actually, millennia, if we count tattoos, piercings, circumcision, and the like.) But can we compare face-lifts and other nips and tucks with an intervention on the genome that may not even be confined to the person it is aimed for? And on that note, do future parents have a right to decide on the physical or intellectual makeup of their progeny? What on earth are parents trying to guarantee or avoid? What is so problematic, for a developing human being, about facing the luck of the draw and defining his or her own destiny by combining willpower with whatever gifts or flaws one is born with? What is so wrong about building character by overcoming bad developmental luck or exercising modesty when one's gifts are favorable? Absolutely nothing, as far as I can see, although a colleague of mine who read this passage complained that I was being too accepting of my flaws—I know, I should have been taller—and that my attitude made me a victim of the Stockholm syndrome, a condition in which hostages become friendly with their captors. I am open to listening to counterarguments and changing my opinion.

There are also important developments in artificial intelligence and robotics, and some of them are also thoroughly inscribed in the homeostatic imperative that governs cultural evolution. Complementing human cognition, from perception and intelligence to motor performance, are old homeostatic-driven practices. Just think of reading glasses, binoculars and microscopes, hearing aids, walking canes, and wheelchairs. Or think, for that matter, of calculators and dictionaries. Artificial organs and prosthetic limbs are not new either, nor are, on the shady side of the street, the performance enhancers that get Olympic athletes and Tour de France champions in so much trouble. Gaining access to strategies and devices that can speed up movement or improve one's intellectual performance is hardly problematic except for competitions.

The application of artificial intelligence to medical diagnostics is very promising. Diagnosis of illnesses and interpretation of diagnostic procedures are the bread and butter of medicine and depend on pattern recognition. Machine learning programs are a natural tool in this area and have achieved reliable and trustworthy results.[2]

By comparison with some of the currently contemplated genetic interventions, the developments in this general area are largely benign and potentially valuable. The most likely and immediate scenario is the achievement of prosthetic enhancement devices that could serve to not only compensate for missing functions but also enhance or augment human perception. Examples include artificial retinal implants for blindness and the development of prosthetic limbs controlled by self-driven mental events, namely, the intention to move a limb. Both examples are a current reality and will be perfected in the near future. They constitute significant entries into the world of human-machine hybridization. Beneficial applications include exoskeletons for victims of accidents who become paraplegics or tetraplegics; exoskeletons are literally second, prosthetic skeletons, set around paralyzed limbs and anchored in the spinal column. These prostheses are moved by computers activated by an outside operator or by the patient. The latter can actually be guided by the patient's *intention* to move, capitalizing on the capture of electric brain signals associated with the will to move.[3] We are well on the way to creating hybrids of living organisms and engineered artifacts, something akin to the cyborgs so beloved of science fiction.

Immortality

Woody Allen once joked that he wanted to achieve immortality by not dying. Little did he know that one day the idea of doing away with death would not be a mere joke. Humans have now figured that the possibility is real, and they have been quietly working toward that goal. And why not? If indeed it would be possible to prolong life indefinitely, should one forgo the option?

The practical answer to this question is clear. It might be worth trying, provided one would not need to confront a supreme creator who might have other plans and provided this forever life could be lived as a good life, without the diseases that become so frequent with prolonged longevity—cancers and the dementias, mostly. The boldness of the project takes your breath away, and so does the arrogance it implies. But once you recover your composure—and weary of falling into the Stockholm syndrome pit again—you say, fine, but let me ask some questions. What are the consequences of such a project, immediately and in the long run, for the individuals and for the societies? What conception of humanity informs the endeavor to make humans eternal?

In terms of basic homeostasis, immortality is perfection, the realization of nature's undreamed dream of life perpetuity. The early conditions of homeostasis were such that they promoted the ongoing life and, unwittingly, life into the future. The unplanned devices ensuring future life included the emergence of genetic machinery. In our futuristic scenario, immortality would be the ultimate stage in the life enterprise, an achievement made all the more intriguing and commendable by the fact that it would arrive by way of human creativity. It appears natural, actually, when one considers that creativity is itself a consequence of homeostasis. But what about the downside? Not all things natural are necessarily good, nor is it advisable to let natural things run unchecked.

Immortality would eliminate the most powerful engine of feeling-driven homeostasis: the discovery that death is inevitable and the anguish that the discovery generates. Should we not worry about the loss of such an engine? Of course we should worry. It can be argued that as backup engines of the process of homeostasis perhaps we might keep pain and suffering, due to causes other than death foretold, and pleasure, too. But would we really? Can one imagine that once our immortality wish would be granted, the radical elimination of pain and suffering would be far behind? And what about pleasure? Would we keep it around and turn the earth into an Eden? Or would we do away with pleasure as well and enter the zombie universe where, I sometimes wonder, some of the paladins of immortality would actually not mind living?

None of this is likely to pass anytime soon, though not for lack of trying by venerable futurists and visionaries. For example, the key idea behind transhumanism is the notion that the human mind can be "downloaded" into a computer, thus guaranteeing its eternal life.[4] At the moment, this is an implausible scenario. It reveals a limited notion of what life really is and also betrays a lack of understanding of the conditions under which real humans construct mental experiences. What the transhumanists would actually be downloading is still a mystery. Not their mental experiences, for certain, at least not if these mental experiences conform to the account most humans would give of their conscious minds and that require the devices and mechanisms I described earlier. One of the key ideas in this book is that minds arise from interactions of bodies and brains, not from brains alone. Are transhumanists planning to download the body, too?

I am open to bold scenarios for the future, and I tend to lament failures of scientific imagination, but I cannot really picture the follow-up to this idea. The essence of the problem is perhaps best explained by indicating why there are clear limits to the application of the notions of code and algorithm—two foundational concepts in computational science and artificial intelligence—to living systems, an issue to which I now turn.

The Algorithmic Account of Humanity

One remarkable development of twentieth-century science is the discovery that both physical structures and the communication of ideas can be assembled on the basis of algorithms that make use of codes. Using an alphabet of nucleic acids, the genetic code helps living organisms assemble the basics of other living organisms and guide their development; likewise, verbal languages provide us with alphabets, with which we can assemble an infinity of words that name an infinity of objects, actions, relationships, and events, and with grammatical rules that govern the sequencing of the words. And so we construct sentences and stories that narrate the course of events or explain ideas. At this point in evolution, many aspects of the assembly of natural organisms and of communication depend on algorithms and on coding, as do many aspects of computation as well as the entire enterprises of artificial intelligence and robotics. But this fact has given rise to the sweeping notion that natural organisms would somehow be reducible to algorithms.

The worlds of artificial intelligence, biology, and even neuroscience are inebriated with this notion. It is acceptable to say, without qualifications, that organisms are algorithms and that bodies and brains are algorithms. This is part of an alleged singularity enabled by the fact that we can write algorithms artificially and connect them with the natural variety, and mix them, so to speak. In this telling, the singularity is not just near: it is here.

The usage and the ideas have gained some currency in technology and science circles and are part of a cultural trend, but they are not scientifically sound. Humanly, they fall short of the mark.

Saying that living organisms are algorithms is in the very least misleading and in strict terms false. Algorithms are formulas, recipes, enumerations of steps in the construction of a particular result. Living organisms, including human organisms, are constructed according to algorithms and use algorithms to operate their genetic machinery. But they are *not* algorithms themselves. Living organisms are consequences of the engagement of algorithms and exhibit properties that might or might not have been specified in the algorithms that directed their construction. Most important, living organisms are collections of tissues, organs, and systems within which every component cell is a vulnerable living entity made of proteins, lipids, and sugars. They are *not* lines of code; they are palpable stuff.

The idea that living organisms are algorithms helps perpetuate the false notion that the substrates used in the construction of an organism, be it living or artificial, are not a relevant issue. It implies that the substrate on which the algorithm operates is not relevant and that neither is the context of the operation. Behind the current usage of the term "algorithm," there seems to lurk the idea of context and substrate independence, although by itself the term does not or should not have such implications.

Presumably, in accordance with current usage, applying the same algorithm to different substrates and in new contexts would achieve similar results. Yet there is no reason why it should be so. Substrates count. The substrate of our life is a particular kind of organized chemistry, a servant to thermodynamics and the imperative of homeostasis. To the best of our knowledge, that substrate is essential to explain who we are. Why is this so? Let me outline three reasons.

First, the phenomenology of feeling reveals that human feelings result from the multidimensional and interactive imaging of our life operations with their chemical and visceral components. Feelings reflect the *quality* of those operations and their future *viability*. Can one imagine feelings arising from a different substrate? Possibly, although there is no reason why such possible feelings would resemble human feelings. I can imagine something "like" feelings arising from an artificial substrate provided they would be reflections of "homeostasis" in the engendered device and would signal the quality and viability of operations in the device. But there is no reason to expect that such feelings would be comparable to those of humans or to those of other species, in the absence of the substrate that feelings actually use to portray the states of living creatures on planet Earth.

I can also envision feelings in a different species somewhere in our galaxy, where life sprang forth and where organisms would have followed a homeostatic imperative similar to ours and generated, on a physiologically different but living substrate, a variant of our feelings. The experience that the mysterious species would have of its feelings would be formally akin to ours, albeit not the same, because the substrate was not exactly the same. If you change the substrate of feelings, you change what gets to be interactively imaged and so you change the feelings as well.

In brief, substrates do count because the mental process to which we are referring is a mental account of those substrates. Phenomenology counts.

There is plenty of evidence that artificial organisms can be designed so as to operate intelligently and even surpass the intelligence of human organisms. But there is no evidence that such artificial organisms, designed for the sole purpose of being intelligent, can generate feelings just because they are behaving intelligently. Natural feelings emerged in evolution, and there they remained because they have made live or die contributions to the organisms lucky enough to have them.

Curiously, pure intellectual processes lend themselves well to an algorithmic account and do not appear to be dependent on the substrate. This is the reason why well-conceived AI programs can beat chess champions, excel at Go, and drive cars successfully. However, there is no evidence to date to suggest that intellectual processes alone can constitute the basis for what makes us distinctly human. On the contrary, intellectual and feeling processes must be functionally interconnected in order to produce something that resembles the operations of living organisms and of humans in particular. Here it is essential to recall the critical distinction, discussed in part II, between emotive processes, which are action programs related to affect, and feelings, the mental experiences of organism states, including the states that result from emotions.

Why is this so important? Because moral values arise out of reward and punishment processes operated by chemical, visceral, and neural processes in creatures equipped with minds. The processes of reward and punishment result in none other than the feelings of pleasure and pain. The values that our cultures have been celebrating in the form of arts, religious beliefs, justice, and fair governance have been forged on the basis of feelings. Once we would remove the current chemical substrate for suffering and for its opposite, pleasure and flourishing, we would remove the natural grounding for the moral systems we currently have.

Of course, artificial systems could be built to operate according to "moral values." That would *not* mean, however, that such devices would contain a grounding for those values and could construct them independently. The presence of "actions" does not guarantee that the organism or device "mentally experiences" the actions.

None of the above implies that the higher, feeling-based functions of living organisms are intangible or are not amenable to scientific investigation. They certainly have been approachable and continue to be. Nor am I arguing against the use of the notion of algorithm for the purpose of introducing mystery in the argument. But until shown otherwise, investigations of living organisms need to take into account the living substrate and the complexity of the resulting processes. The implication of these distinctions is not trivial as we contemplate the new era of medicine we discussed earlier, in which the extension of human life will be possible by means of genetic engineering and the creation of human/artificial hybrids.

Second, the predictability and inflexibility that the term "algorithm" conjures up do not apply to the higher reaches of human behavior and mind. The abundant presence of conscious feeling in humans guarantees that the execution of the natural algorithms can be thwarted by creative intelligence. Our freedom to run against the impulses that either the good or the bad angels of our natures attempt to impose on us is certainly limited, but the fact remains that in many circumstances we can act against such good or bad impulses. The history of human cultures is in good part a narrative of our resistance to natural algorithms by means of inventions not predicted by those algorithms. In other words, even if we were to throw caution to the winds and liberally declare human brains "algorithms," the things that humans do are not algorithms, and we are not necessarily foretold.

One can argue that the departures from natural algorithms are in turn open to an algorithmic account. That is correct, but the point remains that the "initiating" algorithms do not create all the behaviors. Feeling and thinking contribute their share, using their considerable degrees of freedom. If so, then what is the advantage of using the term?

Third, accepting an algorithmic account of humanity that implies the problems outlined above—substrate and context independence, inflexibility, and predictability—is the sort of reductionist position that often leads good souls to dismiss science and technology as demeaning and bemoan the passing of an age in which philosophy, complete with aesthetic sensibility and a humane response to suffering and death, made us soar above the species on whose biological shoulders we were riding. I believe we should not deny the merit of a science project or impede it because it contains a problematic account of humanity. My point is simpler. To produce accounts of humanity that appear to diminish human dignity—even if they are not meant to do so—does not advance the human cause.

Advancing the human cause is hardly the issue for those who believe that we are entering a "post-humanist" phase of history, a phase in which most human individuals have lost their usefulness to society. In the picture painted by Yuval Harari, when humans are no longer required to fight wars—cyber warfare can do that for them—and after humans have lost their jobs to automation, most of them will simply wither away. History will belong to those who will prevail by *acquiring* immortality—or at least long, long longevity—and who will remain to benefit from this arrangement. I say "benefit" rather than "enjoy" because I imagine that the status of their feelings will be murky.[5] The philosopher Nick Bostrom provides another alternative vision, one in which very intelligent and destructive robots will actually take over the world and put an end to human misery.[6] In either case, future lives and minds are presumed to depend at least in part on "electronic algorithms" that artificially simulate what "biochemical algorithms" currently do. Moreover, in the perspective of such thinkers, the discovery that human life is comparable, in its essence, to the life of all other living species undermines the traditional platform of humanism: the idea that humans are exceptional and distinct from other species. This is Harari's apparent conclusion, and if so it is certainly wrong. Humans share numerous aspects of the life process with all other species, but humans are indeed distinct on a number of features. The scope of human suffering and joys is uniquely human, thanks to the resonance of feelings in memories of the past and in the memories they have constructed of the anticipated future.[7] But perhaps Harari just wants to terrify us out of our wits with his *Homo Deus* fable and hopes we will do something about it before it's too late. In that case we agree, and I certainly hope so.

I reproach these dystopian visions on yet another count: they are infinitely colorless and boring. What a comedown from Aldous Huxley's dystopia in *Brave New World*,[8] with its embrace of the pleasurable life. The new visions resemble the repetitive and tedious existence of Luis Buñuel's characters in *The Exterminating Angel*. I much prefer the dangers and the smarts of Alfred Hitchcock's *North by Northwest*. Cary Grant copes with every challenge, outsmarts the arch villain James Mason, and wins Eva Marie Saint.

Robots Serving Humans

Fortunately a good number of the current efforts in the expanding world of AI and robotics are aimed at creating not humanlike robots but devices that *do* things we humans need to have done competently, economically, and faster if possible. The emphasis is on smart action programs. It does not matter at all that the programs do not produce feelings, let alone conscious experiences.[9] I am interested in the "sense" of my robot, not in her "sensibility."

The idea of building humanlike robots that might become our convenient assistants or companions is perfectly reasonable. If artificial intelligence and engineering can take us there, why not? Provided the engineered creatures are under human supervision, provided they have no way of acquiring autonomy and turning against us, and provided we will not have the means to program robots such that they can destroy the world, why not? It must be added that there are several dark scenarios regarding not so much future robots as future AI programs that do have doomsday potential and that need to be watched for. Still, the risks of actual engineered robots going rogue on us is small at this point compared with the real risks of cyber warfare. Do not expect the grandson of HAL, the robot in Stanley Kubrick's *2001: A Space Odyssey*, to show up any day and take over the Pentagon. Expect a number of very bad humans instead.

The reason why such science fiction scenarios are perhaps stronger than ever is the obvious and remarkable successes that intelligent game-playing programs have had in beating champions of chess and Go. The reason why the sci-fi scenarios are not likely to pass is that the sort of intelligence those AI programs exhibit, albeit spectacular, is truly deserving of the name "artificial" and bears limited resemblance to the actual mental processes of human beings. Such AI programs have pure cognition but no affect, which means that the intellectual steps in their "smart" minds cannot indulge in an interplay with prior, accompanying, or predicted feelings. In the absence of feelings, a large part of their hope for humanity vanishes because it is the feeling part of us humans that generates our vulnerabilities, that is essential for us to experience personal suffering and joy and empathize with the suffering and joy of others, and, in sum, is essential to ground a considerable portion of what constitutes morality and justice and assemble the ingredients of human dignity.

When we talk about lifelike, humanlike robots and discover that they do not have feelings, we are talking about an absurd and nonexisting myth. Humans have life and have feelings, and such robots have neither.

Still, the situation can be made more nuanced. We can approximate the process of life in a robot by building into it the conditions of homeostasis that define life to begin with. Although there will be high costs to the robot's efficiency, there is no reason why this cannot be achieved. It consists of engineering a "body" that seeks to satisfy some built-in, homeostasis-like regulatory parameters. The germ for this idea goes back to the pioneer roboticist Grey Walter.[10]

The issue of feelings remains tricky, however. Usually, instead of feeling, roboticists build in toylike behaviors with fake smiles, cries, pouts, and so forth. The result is something like animated emoticons. We are talking about puppetry, really. The actions are not motivated by an internal state of the robot; they are simply programmed into it on the say-so of the designer. They may resemble emotions, in the sense that emotions are action programs, but they are not *motivated* emotions. We still fall for such robots easily, and we are perfectly capable of engaging them as if they were flesh-and-blood creatures. People grow up capable of imagining lives behind the toys and dolls of their early childhood and carry the residue of those identifications. We can easily slide into the world of puppets if the setting is right. In fact, I have never met a robot that I did not like, and they all "seemed" to like me.

If the animations of robots are not emotions, they certainly are not feelings, feelings being, as we know, the mental experience of a body state, which really means subjective mental experiences. And here is when the problem worsens: to have mental experiences, we need minds and not just minds but *conscious* minds. To be conscious, to have subjective experiences, we badly need the two ingredients we described in chapter 9: *an individual perspective of our own organism and individual feeling.* Can we do this in robots? Well, we can in part. I believe we can build perspective in a robot, relatively easily, once we take the problem seriously. But to build feeling, on the other hand, we require a living body. A robot with homeostatic features would be a step in that direction, but the critical issue is the degree to which sketchy body phantoms and some simulation of body physiology could serve as substrates for anything like feeling, let alone human feeling. This is an open and important research question, and we need to investigate it.

Assuming we would make progress in that direction, we might approach the possibility of feeling and, following feeling, of some semblance of humanlike intelligence—in such a context, I can see intuition arising out of Big Data treatments—and a possible entry into humanlike behaviors, complete with predicted risks, felt vulnerabilities, affective attachments, joys, lows, wisdom, the failures and glories of human judgment.

It will not be difficult, even without feelings, for so-called humanlike robots to play and win many sorts of games, or to talk as well as HAL seemed to talk in *2001,* or to serve as helpful human companions, although one shudders a bit at the prospect of a society that *needs* robots as companions. Are there not enough unemployed to fill those jobs after self-driving cars and trucks took away their livelihood? I can see humanlike robots predict the weather, operate heavy machinery, and perhaps even turn against us. But it will take a while until they *really* feel, and until then the simulation of humanity will be just that, a simulation.

Back to Mortality

While we wait for the promised and touted singularities, we might as well deal seriously with two of the largest medical problems anywhere in the world: drug addictions and pain management. The centrality of feelings and homeostasis to an account of human cultures is made very clear by the resistance of these assiduously studied problems to reasonably satisfactory solutions. One can blame drug cartels, big pharma, and irresponsible physicians for ensuring the continuation of drug addictions. They certainly are to blame. We can blame the Internet for making it possible for intelligent and knowledgeable individuals to concoct addictive drugs meshing together otherwise nonaddictive compounds obtained by legal prescriptions. But all that blaming simply misses the point. The addictions are related to molecules that have governed fundamental processes of homeostasis since the mists of time and to an entire suite of opioid receptors. Good, bad, and in-between feelings are tied to what goes on in these receptors, and those feelings, in turn, reflect how well our life is marching, prior to any drug taking. The molecules and receptors on which our feelings depend are old and experienced. They have survived hundreds of millions of years, they are devious, and their effects are powerful. As befits their nature, they produce arresting and tyrannical feelings. The effects of the drugs are destructive to the physical and mental health of the users, achieving the very opposite of the goals of homeostasis. And while people worry about downloading themselves into a computer, these molecules and receptors continue to wreak havoc in the brains and bodies of those people with the misfortune of chronic pain syndromes or a drug addiction, often both.

12

ON THE HUMAN CONDITION NOW

An Ambiguous State of Affairs

Standing at the edge of the Sea of Galilee on a sunny winter morning, down a few steps from the Capernaum synagogue where Jesus of Nazareth talked to his followers, I turn my thoughts from the long gone troubles of the Roman Empire to the current crisis of the human condition. The crisis is intriguing because, although the local conditions across the world are distinct, they elicit similar responses featuring anger and confrontation as well as appeals to isolation and a slide toward autocracy; the crisis is also discouraging because it should not be happening at all. One hoped that at least the most advanced societies had been immunized by the horrors of World War II and the threats of the Cold War and would have found cooperative ways to overcome gradually and peacefully any and all problems that complex cultures face. In retrospect, one should have been less complacent.

This could be the best of times to be alive because we are awash in spectacular scientific discoveries and in technical brilliance that make life ever more comfortable and convenient; because the amount of available knowledge and the ease of access to that knowledge are at an all-time high and so is human interconnectedness at a planetary scale, as measured by actual travel, electronic communication, and international agreements for all sorts of cooperation, in science, the arts, and trade; because the ability to diagnose, manage, and even cure diseases continues to expand and longevity continues to extend so remarkably that human beings born after the year 2000 are likely to live, hopefully well, to an average of at least a hundred. Soon we will be driven around by robotic cars, saving us effort and lives because, at some point, we should have few fatal accidents.

In order to judge our time as the most perfect of times, however, one would need to be quite distracted, not to mention indifferent to the plight of down-and-out fellow humans. Although scientific and technical literacy have never been higher, the public spends little time reading novels or poetry, still the surest and most rewarding way of gaining entry into the comedy and drama of existence and having an opportunity to reflect on who we are or may be. Apparently, there is no time to be spent on the nonpractical matter of just being. A part of the societies that celebrate modern science and technology and that most benefit from them appears to be spiritually bankrupt, in the secular and religious sense of the term spiritual. Judging from their unconcerned acceptance of problematic financial crises—the 2000 Internet bubble, the 2007 mortgage abuses, and the 2008 banking collapse—they appear to be morally bankrupt as well. Intriguingly, or perhaps not so, the level of happiness in the societies that have most gained from the remarkable progress of our time is either stable or declining, assuming we can trust the respective measurements.[1]

For the past four or five decades, the general public of the most advanced societies has accepted with little or no resistance a gradually deformed treatment of news and public affairs designed to fit the entertainment model of commercial television and radio. Not so advanced societies have followed suit with no difficulty. The conversion of nearly all public-interest media into for-profit businesses further reduced the quality of the information. Although a viable society must care for the way its governance promotes the welfare of citizens, the notion that one should pause for some minutes of each day and make an effort to learn about the difficulties and successes of governments and citizenry is not just old-fashioned; it has nearly vanished. As for the notion that we should learn about such matters seriously and with respect, it is by now an alien concept. Radio and television transform every governance issue into "a story," and it is the "form" and entertainment value of the story that count, more than its factual content. When Neil Postman wrote his book *Amusing Ourselves to Death: Public Discourse in the Age of Show Business*, in 1985, he made an accurate diagnosis, but he had no idea that we would suffer so much before dying.[2] The problem has been compounded by the defunding of public education, and the predictable decline of the preparation of citizens, and aggravated, in the case of the United States, by the 1987 repeal of the 1949 Fairness Doctrine that required licensees of the public airwaves to present public affairs in an equitable and honest manner. The result, made more acute by the decline of printed media and the rise and near-complete dominance of digital communication and television, is a profound lack of detailed and nonpartisan knowledge of public affairs coupled with a gradual abandonment of the practices of calm reflection and discernment over facts. One should be careful not to exaggerate the nostalgia for a time that never quite was. Not everyone was seriously informed, reflective, and discerning. Not everyone had reverence for truth and nobility of spirit, not to mention reverence for life. Still, the current breakdown in serious public awareness is problematic. Human societies are predictably fragmented according to an assortment of measures such as literacy, level of education, civic behavior, spiritual aspirations, freedom of speech, access to justice, economic status, health, and environmental safety. Under the circumstances, it is ever more difficult to encourage the public to promote and uphold a shared slate of nonnegotiable values, rights, and obligations for its citizens.

Given the astounding progress of new media, the public has a chance of learning in greater detail than ever about the real facts behind economies, the state of local and global governments, and the state of the societies they live in, which is, no doubt, an empowering advantage; moreover, the Internet provides means of deliberation outside traditional commercial or governmental institutions, another potential advantage. On the other hand, the public generally lacks the time and the method to convert massive amounts of information into sensible and practically usable conclusions. Moreover, the companies that manage the distribution and aggregation of the information assist the public in a dubious way: the flow of information is directed by company algorithms that, in turn, bias the presentation so as to suit a variety of financial, political, and social interests, not to mention the tastes of users so that they can continue within their own entertaining silo of opinions.

One should acknowledge, in fairness, that the voices of wisdom from the past—the voices of experienced and thoughtful editors of newspapers and radio and television programs—were also biased and favored particular views of how societies should function. In several cases, however, those particular views were identifiable with specific philosophical or sociopolitical perspectives, and one could either endorse the conclusions or resist them. The general public has no such opportunity today. Each person has direct access to the world via his or her own fully apped portable device and is encouraged to maximize his or her autonomy. There is little incentive to engage, let alone accommodate, the dissenting views of others.

The new world of communication is a blessing for the citizens of the world trained to think critically and knowledgeable about history. But what about citizens who have been seduced by the world of life as entertainment and commerce? They have been educated, in good part, by a world in which negative emotional provocation is the rule rather than the exception and where the best solutions for a problem have to do primarily with short-term self-interests. Can they really be blamed?

The widespread availability of nearly instantaneous and abundant communication of public and personal information, a manifest benefit, paradoxically reduces the time required for reflection on that same information. The management of the flood of available knowledge often requires a rapid classification of facts as good or bad, likable or not. This possibly contributes to an increase of polarized opinions regarding social and political events. The exhaustion over the flood of facts recommends withdrawal into default beliefs and opinions, often those of the group to which one belongs. This is further aggravated by the fact that no matter how smart and well informed one is, we naturally tend to resist changing our beliefs, in spite of the availability of contrary evidence. Work from our institute demonstrates this point for political beliefs, but I suspect it applies to a wide range of beliefs from religion and justice to aesthetics. The work shows that resistance to change is associated with a conflicting relationship of brain systems related to emotivity and reason. The resistance to change is associated, for example, with the engagement of systems responsible for producing anger.[3] We construct some sort of natural refuge to defend ourselves against contradictory information. Disaffected electorates across the world fail to show up at voting booths. In such a climate, the spreading of false news and post-truths is made easier. The dystopian world that George Orwell once described with the Soviet Union in mind has returned to fit a different sociopolitical situation. Speed of communication and the resulting acceleration of the pace of life are also possible contributors to a decline in civility, detectable in the impatience of public discourse and in the increased rudeness of urban life.[4]

A separate but important issue that continues to be unappreciated is the addictive nature of electronic media, from simple e-mail communications to social networks. The addiction diverts time and attention from the immediate experience of surroundings to a mediated experience via all sorts of electronic devices. The addiction enhances the misfit between the volume of information and the time required to process it.

The breakdown in privacy that accompanies the universal use of the web and of social media guarantees the monitoring of every human move and expressed idea. Moreover, all sorts of surveillance ranging from the necessary for public security to the intrusive and downright abusive are now a reality, practiced by governments and by the private sector, with guaranteed impunity. Surveillance makes espionage, even the espionage of superpowers, a well-established activity that has been with us for millennia, sound honorable and childish. Surveillance is even for sale for high profit from a variety of tech companies. Unfettered access to private information is being used to generate embarrassing scandals, even if the subject matter may not be of a criminal nature. The result is the cowering of political candidates into silence lest they and their political campaigns be destroyed by personal revelations. This has now become another important factor in public governance. In large sectors of the most technically advanced regions of the world, scandals large and small have influenced electoral results and enhanced the public's growing mistrust of political establishments and professional elites. Societies already facing major problems of inequality of wealth and human dislocations due to unemployment and wars have become nearly ungovernable. Disoriented electorates refer to long gone and mythically better pasts with either nostalgia or angry revolt. But the nostalgia is misplaced, and the anger is often misdirected. They reflect a limited understanding of the plethora of facts served by varied media and designed primarily to entertain, promote particular social, political, and commercial interests, and reap huge financial rewards in the process.

There is a growing tension between the power of a large public that appears better informed than ever but does not have the time or the tools to judge and interpret the information and the power of the companies and governments that control the information and know everything there is to be known about that same public. How the resulting conflict can be resolved is not clear.

There are other risks. Catastrophic conflicts involving nuclear and biological weapons pose real and possibly higher risks now than when the weapons were controlled by Cold War powers; the risks of terrorism and the newly added risk of cyber warfare are also real and so is the risk of antibiotic-resistant infections. We may lay the blame for all of these concerns at the feet of modernity, globalization, inequality of wealth, unemployment, too little education, too much entertainment, diversity, and the radically paralyzing speed and ubiquity of digital communication. But the prospect of ungovernable societies remains the same no matter the cause or causes.

This bleak view has been tempered by that of Manuel Castells, one of the foremost scholars of communication technology and a leading sociologist, whose work is essential to understanding power struggles in twenty-first-century cultures. For example, by revealing the inadequacy and corruption of governance in leading democracies, he believes that digital media have actually opened the way for a deep and hopefully healthy remodeling of governance. We have not seen the good results yet. For Castells, a rearrangement of human powers compatible with democracy is still possible. Castells is also skeptical that there ever was a mythical age in which media, education, civic behavior, and governance would have been less problematic than they are today. Liberal democracies have a crisis of legitimacy that needs to be addressed sooner rather than later. The Internet and more generally digital communication have a positive role to play and would be more of a blessing than a curse.[5]

It is important to celebrate the widespread recognition of human rights and the growing attention given to the violation of those rights. The seeds for considering that the core characteristics of human beings are the same anywhere in the world and have roots in a universal common ancestor have been successfully sown. It is more generally accepted than ever that humans are equally entitled to pursue happiness and to have their dignity respected. After World War II, the United Nations adopted the Universal Declaration of Human Rights, the closest we have come to a desirable but so far unwritten international law, conferring the same rights to all humans; violations of those rights, in some parts of the world, can be brought before international tribunals as crimes against humanity. Humans are obligated to other humans and maybe one day they will also be obligated to other living species and to the planet they were born into. This is real progress. The circle of human concerns has definitely enlarged, as Amartya Sen, Onora O'Neill, Martha Nussbaum, Peter Singer, and Steven Pinker, among others, have noted.[6] But why are we witnessing the weakening or collapse of the very establishments that have made these advances possible? Why have things gone wrong, once again, in humanity's progress in ways that disturbingly resemble the past? Can biology help explain why?

Is There a Biology Behind the Cultural Crisis?

What can we say about the meaning of this state of affairs in biological terms? Why is it that humans periodically wipe out the cultural gains they have made, at least in part? Understanding the biological underpinnings of the human cultural mind is not a complete answer but may help us cope with the problem.

In fact, from the biological perspective I have outlined, the repeated failures of cultural efforts should not be surprising. Here is why. The physiological rationale and primary concern of basic homeostasis is the life of an individual organism within its borders. Under the circumstances, basic homeostasis remains a somewhat parochial affair, focused on the temple that human subjectivity has designed and erected—the self. It can be extended with more or less effort to the family and to the small group. It can be extended further out, to larger groups, on the basis of circumstances and negotiations in which the prospects of general benefits and power are well balanced. But homeostasis, as found in each of our individual organisms, is not *spontaneously* concerned with very large groups, especially heterogeneous groups, let alone with cultures or civilizations as a whole. To expect *spontaneous* homeostatic harmony from large and cacophonous human collectives is to expect the unlikely.

Unfortunately, "societies," "cultures," and "civilizations" tend to be regarded as large and singular living organisms. They are conceived, in many respects, as bigger versions of an individual human organism similarly animated by the purpose of persisting and flourishing as a unit. In a metaphoric way, they are so, of course, but in reality that is rarely the case. Societies, cultures, and civilizations are usually fragmented, constituted by juxtaposed and separable "organisms," each with more or less ragged borders. Natural homeostasis tends to do its job relative to *each* separable cultural organism and no more. Left to their own devices, without the countervailing effect of determined civilizational efforts aimed at some degree of integration and the benefit of favorable circumstances, the cultural organisms do not appear to coalesce.

The distinction can be made clearer with an illustration from biology. In our individual human organisms, under normal conditions, the circulatory system does not fight the nervous system for dominance, nor does the heart duel with the lungs to decide on which is more important. But that peaceable arrangement does not apply to social groups within a country, or to countries within a geopolitical union. On the contrary, they frequently engage in battle. Conflicts and struggles for power among social groups are integral components of cultures. Sometimes the conflict may even result from the application of an affect-motivated solution to a prior problem.

The blatant exceptions to the rules that govern the homeostasis of a natural, *individual* organism are grave situations such as malignant cancers and autoimmune diseases; unchecked, they not only fight other parts of the organism to which they belong but can actually achieve organism destruction.

Human groups have made the most sophisticated discoveries of cultural life regulation in different geographic environments and at different points in their respective histories. Diversity of ethnicity and of cultural identity, a fundamental feature of humanity, is the natural outcome of such variety, and it tends to enrich all participants. Diversity, however, contains the germ of conflict. It deepens in-group and out-group fault lines, fosters hostility, and makes general governance solutions more difficult to reach and implement, all the more so in an age of globalization and cross-fertilization of cultures.

The solution to this problem is unlikely to be a forced homogenization of cultures, which in practice is unattainable and undesirable. The idea that homogeneity alone would make societies more governable ignores a biological fact: within the same ethnic group, individuals will differ in terms of affect and temperament. In part, it is likely that such differences are aligned with distinct preferences for certain types of governance and distinct profiles of moral values, as I believe Jonathan Haidt's work implies.[7] The only reasonable and hopefully viable solution for the problem consists of major civilizational efforts through which, by means of education, societies manage to cooperate around fundamental requirements of governance, in spite of differences, large and small.

Nothing short of a massive and enlightened negotiation between affect and reason could ever succeed. But are we guaranteed success if such an extraordinary effort is ever undertaken? I would say the answer is no. There are other sources of disharmony, besides the conflicts generated by the difficulties of reconciling individual interests with the interests of small and large groups. I am referring to *conflicts that originate within each individual,* in the inner clash between positive, loving impulses and negative, hetero and self-destructive impulses. In the last years of his life, Sigmund Freud saw the bestiality of Nazism as confirming his doubts that culture could ever tame the nefarious death wish that he believed was present within each of us. Freud had earlier begun to articulate his reasons in the collection of texts known as *Civilization and Its Discontents* (published in 1930 and revised in 1931),[8] but nowhere are his arguments better expressed than in his correspondence with Albert Einstein. Einstein wrote to Freud in 1932 seeking his advice on how to prevent the deadly conflagration he saw coming, following fast on the heels of World War I. In his reply, Freud described the human condition with merciless clarity and lamented to Einstein that given the forces at play he had no good advice to offer, no help, no solution, I'm so sorry.[9] The main reason for his pessimism, it should be noted, was the internally flawed condition of the human. He did not primarily blame the cultures or specific groups. He blamed the human beings.

Then as now, what Freud called a "death wish" remains an important factor behind human social failures, although I would describe it in less mysterious and poetic words. That factor, as I see it, is a structural component of the human cultural mind. In contemporary neurobiological terms, Freud's "death wish" corresponds to the unrestrained triggering of a specific set of negative emotions, their subsequent disruption of homeostasis, and the overwhelming havoc they cause on individual and collective human behaviors. These emotions are part of the machinery of affect discussed in chapters 7 and 8. We know that several "negative" emotions are actually important protectors of homeostasis. They include sadness and grief, panic and fear, and disgust. Anger is a special case. It has remained in the human emotion tool kit because it can, under certain circumstances, give an advantage to the angry subject by causing the adversary to recoil. But even when it gives advantages anger tends to have high costs, especially when it escalates to ire and violent rage. Anger is a good example of a negative emotion whose benefits have been diminishing in evolution. So are envy, jealousy, and contempt prompted by humiliations and resentments of all sorts. It is commonly said that the engagement of such negative emotions is a return to our animal emotionality, but that is an unnecessary insult to so many animals. The assessment is partly correct but does not begin to capture the bleaker nature of the problem. In humans, the destructiveness of raw greed, anger, and contempt, for example, has been responsible for unthinkable cruelty perpetrated by humans on other humans since prehistoric times. It does resemble, in many ways, the cruelty of our ape cousins, famous for tearing into the bodies of rivals, real or presumed, but it has been made worse by human refinements. Chimpanzees have never crucified other chimpanzees, but Romans invented crucifixion and crucified humans. It takes creative human invention to design new methods for torturing and killing. Human anger and malice are assisted by abundant knowledge, twisted reasoning, and the unbridled powers of technology and science that humans have at their disposal. It does appear that fewer humans today engage in the malicious destruction of others, and that is a sign that some progress has been made. But the potential for mass destruction that those fewer individuals have at their disposal has never been greater. Freud was perhaps struggling with this fact when he asked himself, at the beginning of chapter 7 in *Civilization*, why was it that animals did not have cultural struggles? He did not answer his question, and yet it is clear that animals lack the intellectual apparatus to do so. We do not.

The degree to which the nefarious impulses are present in human societies and how much they influence public behavior is not evenly distributed across populations. To begin with, there are gender differences.[10] Males are still more likely to be physically violent than females in keeping with their ancestral social roles—hunting and fighting for territory—and females can be violent too, but it is apparent that the majority of males are caring individuals and that not all females are. There is plenty of nourishing affect to be found on both sides of the aisle.

Acting on impulses, good or bad, has other constraints. It depends on individual temperament, for example, which in turn depends on how drives and emotions are typically deployed in an individual as a result of numerous factors—genetics, early life development and experience, and historical and social environments, where family structure and education figure prominently. The expression of temperament is even influenced by current social environment and by climate.[11] Cooperative strategies have been a part of the homeostatically driven biological makeup of humans, which means that the germ of conflict resolution is present in human groups, along with the tendency for conflicts. It seems reasonable to assume, however, that the balance between salutary cooperation and destructive competition depends, in substantial part, on civilizational containment and on fair and democratic governance, representative of those who are governed. In turn, civilizational containment depends on knowledge, discernment, and at least a modicum of the wisdom that results from education, scientific and technical progress, and the modulation of humanist traditions, religious as well as secular.

Barring such determined efforts of civilization, groups of individuals with distinct cultural identities and the related psychological, physical, and sociopolitical features will struggle to obtain what they need or want by the available means. This is precisely what the homeostatically driven biological makeup of the groups naturally promotes, once they coalesce as a fuzzy-bordered entity. Other than through the despotic control of one group over another or others, the only way to prevent or resolve destructive struggles is to engage in cooperative behaviors, the sorts of intelligent negotiations of conflict that hallmark human societies at their civilized best.

The mounting of such cooperative efforts also requires the presence of governance leaders accountable to the individuals expected to benefit, along with an educated citizenry that can implement the efforts and monitor the results. I note that it may appear, at first glance, that when we turn to governance, we leave the realm of biology. But that is simply not true. *The protracted negotiating process required for governance efforts is necessarily embedded in the biology of affect, knowledge, reasoning, and decision making. Humans are inevitably caught inside the machinery of affect and its accommodations with reason. There is no exit from that condition.*

Leaving aside the successes of the past, how likely is it that a civilizational effort will succeed today? In one possible scenario, it will not succeed at all, because the very instruments with which we invent cultural solutions—a complicated interplay of feelings and reasons—are undermined by the conflicting homeostatic goals of different constituencies: the individual, the family, the cultural identity group, and larger social organisms. In this version of our predicament, the periodic failure of cultures would be due to the very old and prehuman biological origins of some of our distinctive behavioral and mental features, a sort of unwashable original sin whose features permeate and corrupt the solutions for human conflict as well as their application.

Because current cultural solutions or their application or both would not have gained independence from their biological origins, some of our best and noblest intents would be inevitably thwarted. No amount of transgenerational educational effort would be likely to correct that flaw. We would be repeatedly pulled down in the good tradition of Sisyphus, who, as punishment for his arrogance, was condemned to push a big stone up a hill, only to see it roll down and have to start again.

A sidebar to the failure scenario has been articulated by historians and philosophers versed in the world of AI and robotics.[12] As noted in the previous chapter, they imagine that scientific and technological progress will downgrade the status of humans and humanity; they forecast the emergence of superorganisms; and they predict that neither feelings nor consciousness will have a place in future organisms. The science behind these dystopian visions is open to dispute, and the predictions may be inaccurate. But even if the predictions were to be accurate, I see no reason to acquiesce in this version of the future without resistance.

In another scenario, cooperation eventually comes to dominate thanks to a sustained civilizational endeavor over multiple generations. In many respects, notwithstanding the deadly human catastrophes of the twentieth century, there have been numerous positive developments over human history. After all, we did abolish slavery, a widespread cultural practice for thousands of years, and it is difficult to imagine a sane human being capable of defending the practice today. In the culturally advanced Athens of the Plato, Aristotle, and Epicurus that we so justifiably admire, out of a population of about 150,000, only 30,000 were citizens; the rest were slaves.[13] Vagaries and downturns aside, attention has been paid and advances have been made.

Education, in the broadest sense of the term, is the obvious way forward. A long-term educational project aimed at creating healthy and socially productive environments needs to give prominence to ethical and civic behaviors and encourage classical moral virtues—honesty, kindness, empathy and compassion, gratitude, modesty. It should also address human values that transcend the management of life's immediate needs.

The circle of concerns for other humans and, more recently, the concern for nonhuman species and for the planet reveal a growing recognition of the human plight and even an awareness of the particular conditions of life and environment. Some statistics also indicate a decline in some modalities of violence, although the trends may not be sustainable. In this scenario, the worst part of the barbaric human nature would have already been tamed, and cultures would eventually achieve effective control over barbarism and conflict, if only we give them time, a nice prospect indeed. Culturally, we would be too much of a work in progress, far from conforming, in the sociocultural space, to the homeostatic near perfection that has been achieved at the basic biological level over billions of years of evolution. Given that evolution needed so much time to optimize homeostatic operations, how could one expect, in just the modest thousands of years of our shared human condition, to have harmonized the homeostatic needs of so many and such diverse cultural groups? This scenario accommodates temporary setbacks but holds hope for some progress, in spite of the current crisis of liberal democracies.

It is not the first time that dark and sunny scenarios of human nature are contrasted before our eyes. In the middle of the seventeenth century, the vision we traditionally identify with Thomas Hobbes saw humans as solitary, nasty, and brutish. One century later, the vision of humanity that we commonly attribute to Jean-Jacques Rousseau, on the contrary, saw humans as gentle, noble, and, as their journey commenced, uncorrupted. Although Rousseau eventually recognized that society corrupted the angelic purity of humans, neither vision captured the entire picture.[14] Most humans can actually be brutish, savage, cunning, self-interested, noble, silly, innocent, and lovely. No one manages to be all of it at the same time, although some try. The sunny or dark visions of humanity are still intact in contemporary scholarship. The argument that our awareness of the dignity of human life has increased and that progress is possible, to which I referred earlier, is countered by the reality of periodic failures. This is the position of the philosopher John Gray, an unreformed pessimist, who believes that progress is an illusion, a seductive song invented by those who converted to Enlightenment myths.[15] Enlightenments do have their dark, unilluminated parts, something that Max Horkheimer and Theodor Adorno recognized in the middle of the twentieth century.[16]

Nonetheless, one solid reason for hope in the middle of the current crisis is the fact that, to date, no educational project has been pursued consistently enough, long enough, and widely enough to prove beyond a shadow of a doubt that it would not lead to the better human condition we yearn for.

An Unresolved Clash

Troubled but hopeful or troubled and hopeless, it is not possible to decide which of the two scenarios is most likely to pass. There are simply too many unknowns, and the ultimate consequences of digital communication, artificial intelligence, robotics, and cyber warfare hold an especially wild card. Science and technology can be used with great advantage to enhance our future—their potential remains extraordinary—or they can spell our doom. In the meantime, one's preference for the first or second scenario has a lot to do with one's sunny or dark disposition. The problem is that even one's typical disposition tends to flicker between light and dark when it comes to so much trouble and uncertainty. Meanwhile, we can approach the problem with equanimity and conclude as follows.

The human condition encompasses two worlds. One world is made up of the nature-given rules of life regulation, the strings of which are pulled by the invisible hands of pain and pleasure. We are not conscious of the rules or of their undergirding; we are only conscious of certain outcomes we call pain or pleasure. We had nothing to do with the making of the rules—nor, for that matter, with the existence of the powerful forces of pain and pleasure—and we cannot modify them any more than we can change the movements of the stars or prevent earthquakes. We also had nothing to do with the way natural selection has operated for eons to build the apparatus of affect that in good part governs our social and individual lives on the grounds of limiting pain and enhancing pleasure, largely at the individual level, with only partial consideration for other individuals, even for those who are part of the in-group.

There is, however, another world. We could and did work around the conditions imposed on us by inventing cultural forms of life management to complement the basic variety. The result was the discoveries we continue making about universes within and around us and our extraordinary ability to accumulate knowledge in both internal memory and external records. Here the situation is different. We can reflect on the knowledge, think through it, manipulate it intelligently, and invent all sorts of responses to nature's rules. On occasion, our knowledge, which includes, ironically, the discovery of life regulation rules that we cannot modify, lets us do something about the cards we have been dealt. Cultures and civilizations are the names we give to the cumulative results of these efforts.

It has been so difficult to manage the gulf between the naturally imposed life regulation and the responses we invent that the human condition has often resembled a tragedy and, perhaps not often enough, a comedy. The ability to invent solutions is an immense privilege but prone to failure and quite costly. We can call this the burden of freedom or, more precisely, the burden of consciousness.[17] Had we not *known* of the condition—had we not *subjectively felt* it—we would not have cared. But once our subjectively driven *care* took charge of responding to our condition, we biased the process toward our understandable individual interests, which, left to themselves, include the circle of those nearest to us and barely extend to our cultural group. The move has undermined our efforts, at least in part, and actually disrupted homeostasis at different points of a global cultural system.

mandments on a mountain; when, in the name of Buddha, they conceived Nirvana; when, under the guise of Confucius, they came up with ethical precepts; and when, in the roles of Plato and Aristotle and Epicurus, they began explaining to fellow Athenians within earshot how the good life could be lived. Their job was never finished.

A life not felt would have needed no cure. A life felt but not examined would not have been curable. Feelings launched and have helped navigate a thousand intellectual ships.

13

THE STRANGE ORDER OF THINGS

The title of this book was suggested by two facts. The first is that as early as 100 million years ago some species of insects developed a collection of social behaviors, practices, and instruments that can appropriately be called cultural when we compare them with the human social counterparts. The second fact is that even further back in time, in all likelihood several billion years ago, unicellular organisms also exhibited social behaviors whose schematics conform to aspects of human sociocultural behaviors.

These facts certainly contradict a conventional notion: that something as complex as social behaviors capable of improving life management could only have sprung from the minds of evolved organisms, not necessarily human, but complex enough and close enough to humans to engender the requisite sophistication. The social features that I am writing about emerged early in the history of life, are abundant in the biosphere, and did not have to wait for anything human-like to show up on Earth. This order is strange indeed, unexpected to say the least.

A closer look reveals details behind these intriguing facts, for example, successful cooperative behaviors of the sort that we tend to associate, and reasonably so, with human wisdom and maturity. But cooperative strategies did not have to wait for wise and mature minds to appear. Such strategies are possibly as old as life itself and were never more brilliantly displayed than in the convenient treaty celebrated between two bacteria: a pushy, upstart bacterium that wanted to take over a bigger and more established one. The battle resulted in a draw, and the pushy bacterium became a cooperative satellite of the established one. Eukaryotes, cells with a nucleus and complicated organelles such as mitochondria, were probably born this way, over the negotiating table of life.

The bacteria in the above tale do not have minds, let alone wise minds. The pushy bacterium operates *as if* concluding that "when we cannot win over them, we might as well join them." The established bacterium, on the other side, operates as if thinking, "I may as well accept this invader provided it offers something to me." But neither bacterium *thought* anything, of course. No mental reflection was involved, no overt consideration of prior knowledge, no cunning, guile, kindness, fair play, or diplomatic conciliation. The equation of the problem was resolved blindly and from *within* the process, bottom up, as an option that, in retrospect, worked for both sides. The successful option was shaped by the imperative requirements of homeostasis, and that was not magic, except in a poetic sense. It consisted of concrete physical and chemical constraints applied to the life process, within the cells, in the context of their physicochemical relations with the environment. Of note, the idea of algorithm is applicable to this situation. The genetic machinery of the successful organisms made sure the strategy would remain in the repertoire of future generations. Had the option not worked, it would have joined the large graveyard of evolution. We would never have known that fact.

The intriguing process of cooperation does not stand alone, unaided. Bacteria are able to sense the presence of others thanks to the chemical probes installed in their membranes, and they can even tell relatives from strangers via the molecular structure of those probes. This is a modest forerunner of our sensory perceptions, closer to taste and smell than to the image-based hearing or seeing.

These strangely ordered emergences reveal the deep power of homeostasis. The indomitable imperative of homeostasis operated by trial and error to select naturally available behavioral solutions to a number of problems of life management. The organisms searched and screened, unwittingly, the physics of their environments and the chemistry within their walls and came up, unwittingly, with at least adequate but often good solutions for the maintenance and flourishing of life. The marvel is that when comparable problem configurations were encountered on other occasions, at other points in the messy evolution of life-forms, the same solutions were found. The tendency toward particular solutions, toward similar schemes, toward some degree of inevitability, results from the structure and circumstances of living organisms and their relation to the environment and depends on homeostasis writ large. All of which puts one in mind of D'Arcy Wentworth Thompson's writings on growth and form—for example, the forms and structures of cells, tissues, eggs, shells, and so forth.[1]

Cooperation evolved as a twin to competition, which helped select the organisms that exhibited the most productive strategies. As a consequence, when we behave cooperatively today, at some personal sacrifice, and when we call that behavior altruistic, it is not the case that we humans have invented the cooperative strategy out of the kindness of our hearts. The strategy emerged strangely early and is now old hat. What is certainly different and "modern" is the fact that when we encounter a problem that can be resolved with or without an altruistic response, we now can think and *feel* through the process in our minds and can, at least in part, deliberately select the approach we will deploy. We have options. We can affirm altruism and suffer the attending losses or withhold altruism and not lose anything, or even gain, at least for a while.

The issue of altruism is through and through a good entry into the distinction between early "cultures" and the full-fledged variety. The origin of altruism is blind cooperation, but altruism can be deconstructed and taught in families and schools as a deliberate human strategy. As is the case with several benevolent and beneficent emotions—compassion, admiration, awe, gratitude—altruistic behavior can be encouraged, exercised, trained, and practiced in society. Or not. Nothing guarantees that it will always work, but it is there as a conscious human resource available via education.

Another example of the contrast between origins and fully developed cultures can be seen around the notion of profit. Cells have literally been looking for profit for a very long time, by which I mean governing their metabolism so that it yields positive energy balances. Those cells that really succeed in life are good at generating positive energy balances, that is, "profits." But the fact that profit is natural and generally beneficial does not make it *necessarily* good, culturally speaking. Cultures can decide when natural things are good—and determine the degree of goodness—and when they are not. Greed is just as natural as profit but is not culturally good, contrary to what Gordon Gekko famously affirmed.[2]

The most strangely ordered emergences of high faculties are probably feeling and consciousness. It is not unreasonable—just incorrect—to imagine that the mental refinement we know as feelings would have arisen from the most advanced creatures in evolution, if not from humans alone. The same applies to consciousness. Subjectivity, the hallmark of consciousness, is the ability to own one's mental experiences and endow those experiences with an individual perspective. The prevalent view is still that subjectivity is unlikely to have emerged in any creature besides sophisticated humans. Even more incorrectly, it is frequently assumed that such refined processes as feeling and consciousness must result from the operation of the most modern, most humanly evolved structures of the central nervous system, namely, the glorious cerebral cortices. The public interested in such matters actually favors the cerebral cortices, period, and so do notable neuro-scientists and philosophers of mind. The search for the "neural correlates of consciousness" actively pursued by contemporary scientists has centered on the cerebral cortex exclusively. Not only that, it has focused on the process of vision. Vision is also the process elected by philosophers of mind to ground their discussions of mental experience, subjectivity, and the reference to qualia.

The prevalent view, however, is wrong on all counts. Feelings and subjectivity, as far as we can gather, depended on the prior emergence of nervous systems with central components, but there is no justification for favoring the cerebral cortex as responsible for the job. On the contrary, brain-stem nuclei and nuclei in the telencephalon, all located below the cerebral cortex, are the critical structures to support feeling and, by extension, the qualia that are part of our understanding of consciousness. As far as consciousness is concerned, only two of the critical processes I discussed—the construction of body phantom perspective and the process of integrating experiences—are likely to depend mostly on cerebral cortices. Moreover, the emergence of feeling and subjectivity is not recent at all, let alone exclusively human. It is likely to have happened long ago, over the Cambrian period. Not only are all vertebrates likely to be conscious experiencers of a variety of feelings but so are a number of invertebrates whose central nervous system design resembles that of humans as far as spinal cord and brain stem are concerned. Social insects are likely to qualify, and so do charming octopuses drawing on a very different brain design.

The inescapable conclusion is that feeling and subjectivity are old abilities and that they did not depend on the sophisticated cerebral cortex of upper vertebrates, let alone humans, to make their debut. This qualifies as strange, but once again things get even stranger. Far earlier than the Cambrian period, unicellular organisms could respond to injuries to their integrity with defensive and stabilizing chemical and physical reactions, the latter something akin to flinching and wincing. Well, those reactions are, in practical terms, emotive responses, the sorts of action programs that later in evolution could be represented mentally as a feeling. Curiously, even the process of perspective taking is likely to have a very old origin. The sensing and responding of a single cell have an implicit "perspective," the perspective of that particular "individual" organism and that organism alone, except that the implicit perspective is not secondarily represented in a separate map. That may well be an ancestor to subjectivity, an ancestor that one day did become explicit in organisms with minds. I should insist that brilliant as these early processes are, they are through and through about *behaviors*, smart, useful actions. As far as I can see, there is nothing mental or experiential about them—no mind, no feeling, no consciousness. I am very open to more revelations from the world of very small organisms, but I do not expect to read about the phenomenology of microorganisms anytime soon. Or ever.[3]

In brief, the assembly of what became feelings and consciousness for us was made gradually, incrementally, but *irregularly*, along separate lines of evolutionary history. The fact that we can find so much in common in the social and affective behaviors of single-celled organisms, sponges and hydras, cephalopods, and mammals suggests a common root for the problems of life regulation in different creatures and also a shared solution: obeying the homeostatic imperative.

Looming large in the history of homeostatically satisfying *accretions* is the emergence of nervous systems. Nervous systems opened the way for maps and images, for configurational, "resemblative" representations, and that was, in the deepest of senses, *transformative*. Nervous systems were transformative even if they did not and do not work alone, even if they are primarily servants of a larger calling: maintaining productive, homeostasis-abiding lives in complicated organisms.

The above considerations take us to another important part of the strangely ordered emergence of mind, feeling, and consciousness, one that is subtle and easy to miss. It has to do with the notion that *neither parts of nervous systems nor whole brains are the sole manufacturers and providers of mental phenomena.* It is unlikely that neural phenomena alone could produce the functional background required for so many aspects of minds, but it is certainly the case that they could *not* do so in regard to feelings. A close two-way interaction between nervous systems and the non-nervous structures of organisms is a requirement. Neural and non-neural structures and processes are not just contiguous but *continuous* partners, interactively. They are not aloof entities signaling each other like chips in a cell phone. In plain talk, brains and bodies are in the same mind-enabling soup.

Countless problems of philosophy and psychology can begin to be approached productively once the relationships of "body and brain" are placed in this new light. The entrenched dualism that began in Athens, was grandfathered by Descartes, resisted Spinoza's broadside, and has been fiercely exploited by the computational sciences is a position whose time has passed. A new, biologically integrated position is now required.

Nothing could be more different from the conception of the relation between minds and brains with which I started my career. I began reading Warren McCulloch, Norbert Wiener, and Claude Shannon when I was twenty, and due to several quirks of destiny McCulloch would soon become my first American mentor along with Norman Geschwind. This was a foundational, exhilarating time for science, one that opened the way for the extraordinary successes of neurobiology, the computational sciences, and artificial intelligence. In retrospect, however, it had little to offer by way of a realistic view of what human minds look and feel like. How could it, given that the respective theory disengaged the dried-up mathematical description of the activity of neurons from the thermodynamics of the life process? Boolean algebra has its limits when it comes to making minds.[4]

Something that made good use of the cerebral cortices, although it did not have to wait for cerebral cortices to appear, human or otherwise, was the ability to survey the operations of numerous systems inside living organisms and formulate predictions about the future of those operations, based on the past history of the organism and on its current performance. I am talking, in other words, about surveillance, and I use the term advisedly.

When I described the structure and function of our peripheral nervous systems, I mentioned that given the astonishing continuity and interactivity of nervous systems and organisms, nerve fibers get to "visit" every part of our bodies and report on the local state of operations at all those sites to spinal ganglia, to trigeminal ganglia, and to central nervous system nuclei. In brief, in some sense, nerve fibers are "surveyors" of the organism's vast estates. So are, by the way, the lymphocytes of the immune system that patrol the entire landmass of our bodies in search of the bacterial and viral interlopers that need to be kept at bay. A number of nuclei in the spinal cord, brain stem, and hypothalamus contain the neural know-how required to respond to the information so gathered and act on its basis, defensively, as needed. Moreover, the cerebral cortices can scrutinize reams of prior related data and predict what may happen next. Usefully, they can even anticipate untoward drifts of internal function. The useful predictions are revealed as feelings that are, as we have seen, complex mental experiences that result from blending live data sets originating from certain regions, or even globally, relative to the entire body.[5]

Recently, it has become fashionable in the computer sciences and in the world of artificial intelligence to talk about Big Data and about its predictive powers, as inventions of modern technology. But brains, as noted above, and not human brains alone, have long been "Big Data" handlers when they operate homeostasis at high neural levels. When, for example, we humans intuit the outcome of a particular dispute, we make ample use of our "Big Data" support systems. We draw on past surveillance, recorded in memory, and on prediction algorithms.

It should be noted that the extraordinary surveillance and espionage capabilities of modern governments, social media behemoths, and companies that spy for hire are only the latest users of nature's original unpaid franchise. We cannot fault nature for developing homeostatically useful surveillance systems, on the contrary, but we can question and judge the governments and companies that reinvented the surveillance formula merely to strengthen their power and their monetary worth. Questioning and judging are the legitimate business of cultures.

The ordering of all these culture-related emergences has been strange indeed, hardly fitting one's first guess. There are, however, some welcome exceptions. One would expect philosophical inquiry, religious beliefs, true moral systems, and the arts to have emerged late in evolution and be prevalently human. And so they did, and so they are.

The picture that comes to view when such strangely ordered emergences are considered is now clearer. For most of the history of life, specifically for about 3.5 or more billions of years, numerous species of animals and plants exhibited abundant abilities to sense and respond to the world around, exhibited intelligent social behaviors, and accumulated biological devices that made them live more efficiently or longer or both and made them able to pass on to progenies the secret of their flourishing lives. Their lives exhibited *only the precursors* to minds and feeling and thinking and consciousness but not those faculties themselves.

Missing was the ability to represent a resemblance of the objects and events of reality, both external to the organism *and* internal to it. The conditions for the world of images and minds to materialize began to emerge about half a billion years ago, and human minds appeared even more recently, possibly a mere few hundred thousand years.

The onset of early analogue-form representations permitted the rise of images based on varied sensory modalities and made way for feeling and consciousness. Later, symbolic representations came to include codes and grammars, and the way was clear for the languages of words and math. The worlds of image-based memory, imagination, reflection, inquiry, discernment, and creativity came next. Cultures were their prime manifestations.

Our current lives and their cultural objects and practices can be linked, albeit not easily, to the lives of yore, before there were feelings and subjectivity, before there were words and decisions. The connection between the two sets of phenomena travels in a complicated labyrinth where it is easy to make the wrong turn and get lost. Here and there one can find a guiding thread—Ariadne's thread, that is. The task of biology, psychology, and philosophy is to make the thread continuous.

It is often feared that greater knowledge of biology reduces complex, minded, and willful cultural life to automated, pre-mental life. I believe that is not the case. First, greater knowledge of biology actually achieves something spectacularly different: a deepening of the connection between cultures and the life process. Second, the wealth and originality of so many aspects of cultures are not reduced. Third, greater knowledge about life and about the substrates and processes we share with other living beings does not diminish the biological distinctiveness of humans. It is worth repeating that the exceptional status of humans, over and above everything else they share with other creatures is not in question, and derives from the unique way in which their sufferings and their joys are amplified by individual and collective memories of the past and the imagination of a possible future. Increasing knowledge of biology, from molecules to systems, reinforces the humanist project.

It is also worth repeating that there is no conflict at all between accounts of current human behavior that favor autonomous cultural influence or the influence of natural selection conveyed genetically. Both influences play their parts, in different proportions and order.

Although this chapter is dedicated to reordering the emergence of abilities and faculties that can help us explain our humanity, I have used conventional biology and conventional evolutionary thinking to account for the unexpected strangeness of the revised course of events and for the phenomena that I am trying to explain less conventionally, such as mind, feeling, and consciousness. It is perhaps appropriate, in this context, to make two additional comments.

First, it is quite natural, under the sway of new and powerful scientific findings, to fall for premature certainties and interpretations that time will discard mercilessly. I am prepared to defend my current views on the biology of feelings, consciousness, and the roots of the cultural mind, but I am aware that those views may need to be revised before too long. Second, it is apparent that we can talk with some confidence about the traits and operations of living organisms and of their evolution and that we can locate the beginnings of the respective universe about thirteen billion years ago. We do not have, however, any satisfactory scientific account of the origins and meaning of the universe, in brief, no theory of everything that concerns us. This is a sobering reminder of how modest and tentative our efforts are and of how open we need to be as we confront what we do not know.

ACKNOWLEDGMENTS

The development of a book is a long process of planning and reflecting, but the day does come when one needs to sit down and write. I tend to remember vividly when that happens, for each book, and what were the circumstances. I also tend to return to such memories as if they revealed the key on which the text should be written. In the case of this book, it happened in Provence, at the home of our friends Laura and Emanuel Ungaro, and it followed a conversation with Emanuel on the issue of how specific wounds are often prompts for one's creations. We were talking about a curious book (*L'Atelier d'Alberto Giacometti*) written by Jean Genet, a book that Picasso considered the best ever written about artistic creation. Genet's words—"Beauty has no other origin but the singular wound, different for each person, hidden or visible"—connected well with the idea that feeling is a key player in the cultural process. Now writing could begin in earnest and one year later, in the very same surroundings, I recall explaining the first draft to another friend, Jean-Baptiste Huynh.

I wrote early sections of the book elsewhere in France, at the home of Barbara Guggenheim and Bert Fields. I thank all these friends for the inspiration that they and the places they have invented provide so naturally.

This is also the place to mention a disclaimer on the book's title. On first hearing it, several people have asked me if it refers to Michel Foucault. It certainly does not although I know why they ask: Foucault wrote a book whose original French title is *Les Mots et les Choses*

(*The Words and the Things*), which became, in its English version, *The Order of Things*. Nothing to do with my title whatsoever.

My intellectual home is the Dornsife College of Letters, Arts and Sciences at the University of Southern California. A number of colleagues at our Brain and Creativity Institute were patient enough to read the entire manuscript and discuss several passages in detail. I gained a lot from their comments and I thank them all deeply, but none more than Kingson Man, Max Henning, Gil Carvalho, and Jonas Kaplan. Others whose readings, comments, and encouragement were important are Morteza Dehghani, Assal Habibi, Mary Helen Immordino-Yang, John Monterosso, Rael Cahn, Helder Araujo, and Matthew Sachs.

Another group of colleagues, representing a wider roster of disciplines, was just as generous and made many valuable suggestions. They are Manuel Castells, an exceptional scholar who has accompanied the development of my ideas for several years; Steve Finkel; Marco Verweij; Mark Johnson; Ralph Adolphs; Camelo Castillo; Jacob Soll; and Charles McKenna. I thank them for their exceptional scholarship and intelligent advice.

Still another group kindly read parts of the manuscript or helped answer specific questions. They are Keith Baverstock, Freeman Dyson, Margaret Levi, Rose McDermott, Howard Gardner, Jane Isay, and Maria de Sousa.

Finally, some very patient friends read and commented on versions of the book and listened to my musings on the always vexing issue of preparing epigraphs. They are Jorie Graham, Peter Sacks, Peter Brook, Yo-Yo Ma, and Bennett Miller.

The research on which so much of this book is based has been possible only because of the support of two foundations: The Mathers Foundation, which has for decades been exemplary in backing research in biology, and the Berggruen Foundation, whose president, Nicolas

Berggruen, is unendingly curious about human affairs. I thank both foundations for their trust.

Dan Frank, at Pantheon, is a learned, wise, and disarmingly calm voice, the person you need by your side when you come to a fork in the road and cannot take both options. My gratitude is heartfelt. I also thank Betsy Sallee, in his office, for her attentive help.

Michael Carlisle has been a close friend for over thirty years and my agent for about twenty-five. He is a consummate professional and has a heart. I thank him and his team at Inkwell, especially Alexis Hurley.

I owe a debt of gratitude to Denise Nakamura, whose attention to detail, reliability, and patience are a model, and to Cinthya Nunez who makes the Administrative Office of the Brain and Creativity Institute run smoothly and is always ready to take on a problem at a moment's notice. The manuscript owes a lot to their dedication. I also thank Ryan Veiga, who typed portions of the manuscript and assisted me with the preparation of the bibliography.

Last, I need to say that Hanna reads everything I write and is my best—by which I mean worst—critic. She contributes at every step of the way and in every way imaginable. I always try to convince her to be a coauthor, but to no avail. The largest share of thanks go to her, of course.

NOTES AND REFERENCES

1 ON THE HUMAN CONDITION

1. This statement does not apply to the nonstandard situations of manic or depressive states in which feelings may no longer be accurate indicators of the homeostatic state.

2. To read more on affect—drives, motivations, emotions, and feelings, turn to chapters 7 and 8. For other relevant work, turn to Antonio Damasio, *Descartes' Error* (1994; New York: Penguin Books, 2010); Antonio Damasio, *The Feeling of What Happens: Body and Emotion in the Making of Consciousness* (New York: Harcourt, 1999); Antonio Damasio and Gil B. Carvalho, "The Nature of Feelings: Evolutionary and Neurobiological Origins," *Nature Reviews Neuroscience* 14, no. 2 (2013): 143–52; Jaak Panksepp, *Affective Neuroscience: The Foundations* (New York: Oxford University Press, 1998); Jaak Panksepp and Lucy Biven, *The Archaeology of Mind* (New York: W. W. Norton, 2012); Joseph Le Doux. *The Emotional Brain* (New York: Simon & Schuster, 1996); Arthur D. Craig, "How Do You Feel? Interoception: The Sense of the Physiological Condition of the Body," *Nature Reviews Neuroscience* 3, no. 8 (2002): 655–66; Ralph Adolphs, Daniel Tranel, Hanna Damasio, and Antonio Damasio, "Impaired Recognition of Emotion in Facial Expressions Following Bilateral Damage to the Human Amygdala," *Nature* 372, no. 6507 (1994): 669–72; Ralph Adolphs, Daniel Tranel Hanna Damasio, and Antonio Damasio, "Fear and the Human Amygdala," *Journal of Neuroscience* 15, no. 9 (1995): 5879–91; Ralph

Adolphs, Daniel Tranel, Antonio Damasio, "The Human Amygdala in Social Judgment," *Nature* 393, no. 6684 (1998); Ralph Adolphs, F. Gosselin, T. Buchanan, Daniel Tranel, P. Schyns, and Antonio Damasio, "A Mechanism for Impaired Fear Recognition After Amygdala Damage," *Nature* 433, no. 7021, (2005): 68–72; Stephen W. Porges: *The Polyvagal Theory* (New York and London: W. W. Norton, 2011); Kent Berridge & Morten Kringelbach, *Pleasures of the Brain* (Oxford: Oxford University Press, 2009); Mark Solms, *The Feeling Brain: Selected Papers on Neuropsychoanalysis* (London: Karnac Books, 2015); Lisa Feldman Barrett, "Emotions Are Real," *Emotion* 12, no. 3 (2012): 413.

3. This date continues to be revised backward to perhaps as early as 400,000 years ago in the case of the Iberian Peninsula. Richard Leakey, *The Origin of Humankind* (New York: Basic Books, 1994); Merlin Donald, *Origins of the Modern Mind: Three Stages in the Evolution of Culture and Cognition* (Cambridge, Mass.: Harvard University Press, 1991); Steven Mithen, *The Singing Neanderthals: The Origins of Music, Language, Mind, and Body* (Cambridge, Mass.: Harvard University Press, 2006); Ian Tattersall, *The Monkey in the Mirror: Essays on the Science of What Makes Us Human* (New York: Harcourt, 2002); John Allen, *Home: How Habitat Made Us Human* (New York: Basic Books, 2015); Craig Stanford, John S. Allen, and Susan C. Anton, *Exploring Biological Anthropology: The Essentials* (Upper Saddle River, N.J.: Pearson, 2012). CARTA, the Center for Academic Research and Training in Anthropogeny, is a provider of first-rate scientific information on the investigation of the origin of humans, a field known as anthropogeny. See https://carta.anthropogeny.org/about/carta.

4. Michael Tomasello, *The Cultural Origins of Human Cognition* (Cambridge, Mass.: Harvard University Press, 1999); Michael Tomasello, *A Natural History of Human Thinking* (Cambridge, Mass.: Harvard University Press, 2014); Michael Tomasello, *A Natural History of Human Morality* (Cambridge, Mass.: Harvard University Press, 2016).

5. Reports from the London Zoo on the visits of Queen Victoria, in 1842; Jonathan Weiner, "Darwin at the Zoo," *Scientific American* 295, no. 6 (2006): 114–19.

6. The literature consulted for this section includes Paul B. Rainey and Katrina Rainey, "Evolution of Cooperation and Conflict in Experimental Bacterial Populations," *Nature* 425, no. 6953 (2003): 72–74; Kenneth H. Nealson and J. Woodland Hastings, "Quorum Sensing on a Global Scale: Massive Numbers of Bioluminescent Bacteria Make Milky Seas," *Applied and Environmental Microbiology* 72, no. 4 (2006): 2295–97; Stephen P. Diggle, Ashleigh S. Griffin, Genevieve S. Campbell, and Stuart A. West, "Cooperation and Conflict in Quorum-Sensing Bacterial Populations," *Nature* 450, no. 7168 (2007): 411–14; Lucas R. Hoffman, David A. D'Argenio, Michael J. MacCoss, Zhaoying Zhang, Roger A. Jones, and Samuel I. Miller, "Aminoglycoside Antibiotics Induce Bacterial Biofilm Formation," *Nature* 436, no. 7054 (2005): 1171–75; Ivan Erill, Susana Campoy, and Jordi Barbé, "Aeons of Distress: An Evolutionary Perspective on the Bacterial SOS Response," *FEMS Microbiology Reviews* 31, no. 6 (2007): 637–56; Delphine Icard-Arcizet, Olivier Cardoso, Alain Richert, and Sylvie Hénon, "Cell Stiffening in Response to External Stress Is Correlated to Actin Recruitment," *Biophysical Journal* 94, no. 7 (2008): 2906–13; Vanessa Sperandio, Alfredo G. Torres, Bruce Jarvis, James P. Nataro, and James B. Kaper, "Bacteria-Host Communication: The Language of Hormones," *Proceedings of the National Academy of Sciences* 100, no. 15 (2003): 8951–56; Robert K. Naviaux, "Metabolic Features of the Cell Danger Response," *Mitochondrion* 16 (2014): 7–17; Daniel B. Kearns, "A Field Guide to Bacterial Swarming Motility," *Nature Reviews Microbiology* 8, no. 9 (2010): 634–44; Alexandre Persat, Carey D. Nadell, Minyoung Kevin Kim, Francois Ingremeau, Albert Siryaporn, Knut Drescher, Ned S. Wingreen, Bonnie L. Bassler, Zemer Gitai, and Howard A. Stone, "The Mechanical World of Bacteria," *Cell* 161, no. 5 (2015): 988–97; David T. Hughes and Vanessa Sperandio,

"Inter-kingdom Signaling: Communication Between Bacteria and Their Hosts," *Nature Reviews Microbiology* 6, no. 2 (2008): 111–20; Thibaut Brunet and Detlev Arendt, "From Damage Response to Action Potentials: Early Evolution of Neural and Contractile Modules in Stem Eukaryotes," *Philosophical Transactions of the Royal Society B* 371, no. 1685 (2016): 20150043; Laurent Keller and Michael G. Surette, "Communication in Bacteria: An Ecological and Evolutionary Perspective," *Nature Reviews* 4 (2006): 249–58.

7. Alexandre Jousset, Nico Eisenhauer, Eva Materne, and Stefan Sche, "Evolutionary History Predicts the Stability of Cooperation in Microbial Communities," *Nature Communications* 4 (2013).

8. Karin E. Kram and Steven E. Finkel, "Culture Volume and Vessel Affect Long-Term Survival, Mutation Frequency, and Oxidative Stress of *Escherichia coli*," *Applied and Environmental Microbiology* 80, no. 5 (2014): 1732–38; Karin E. Kram and Steven E. Finkel, "Rich Medium Composition Affects *Escherichia coli* Survival, Glycation, and Mutation Frequency During Long-Term Batch Culture," *Applied and Environmental Microbiology* 81, no. 13 (2015): 4442–50.

9. Pierre Louis Moreau de Maupertuis, "Accord des différentes lois de la nature qui avaient jusqu'ici paru incompatibles," *Mémoires de l'Académie des Sciences* (1744): 417–26; Richard Feynman, "The Principle of Least Action," in *The Feynman Lectures on Physics: Volume II*, chap. 19, accessed Jan. 20, 2017, http://www.feynmanlectures.caltech.edu/II_toc.html.

10. Edward O. Wilson has written extensively on the complex social life of insects. His book *The Social Conquest of the Earth* (New York: Liveright, 2012) provides an overview of this spectacular field of research.

11. As noted earlier, the consistent relationship between feelings and homeostasis breaks down during intense negative feelings. Extreme sadness does not necessarily express an extreme deficiency of

basic homeostasis, although it can result in it and even be responsible for suicide. Situational sadness and depression do express unfavorable social situations, and under such circumstances feelings do operate as indicators of danger ahead for homeostatic regulation.

12. Talcott Parsons, "Evolutionary Universals in Society," *American Sociological Review* 29, no. 3 (1964): 339–57; Talcott Parsons, "Social Systems and the Evolution of Action Theory," *Ethics* 90, no. 4 (1980): 608–11. The ideas of other thinkers in the social sciences—such as Pierre Bourdieu, Michel Foucault, and Alain Touraine, are easier to translate into my biological perspective.

13. F. Scott Fitzgerald, *The Great Gatsby* (New York: Scribner's, 1925).

2 IN A REGION OF UNLIKENESS

1. The phrase "region of unlikeness" appears in Saint Augustine, and the poet Jorie Graham used it as the title to one of her first books. For me, it captures the idea that life occurs within a set-aside cellular perimeter and that the process is unlike any other.

2. Freeman Dyson, *Origins of Life* (New York: Cambridge University Press, 1999).

3. Maupertuis, "Accord des différentes lois de la nature qui avaient jusqu'ici paru incompatibles"; Feynman, "Principle of Least Action."

4. See Antonio Damasio, *Looking for Spinoza: Joy, Sorrow, and the Feeling Brain* (New York: Harcourt, 2003).

5. The phrase is the title of a 1946 book by Paul Éluard illustrated by Marc Chagall. William Faulkner, 1949 Nobel Prize acceptance speech, delivered in 1950.

6. Christian de Duve, *Vital Dust: The Origin and Evolution of Life on Earth* (New York: Basic Books, 1995); Christian de Duve, *Singularities: Landmarks in the Pathways of Life* (Cambridge, U.K.: Cambridge University Press, 2005).

7. Francis Crick, *Life Itself: Its Origins and Nature* (New York: Simon & Schuster, 1981).

8. Tibor Gánti, *The Principles of Life* (New York: Oxford University Press, 2003).

9. Richard Dawkins, *The Selfish Gene* (New York: Oxford University Press, 2006).

10. Stanley L. Miller, "A Production of Amino Acids Under Possible Primitive Earth Conditions," *Science* 117, no. 3046 (1953): 528–29.

11. In addition to the work cited earlier, the literature consulted in the preparation of this text includes Eörs Szathmáry and John Maynard Smith, "The Major Evolutionary Transitions," *Nature* 374, no. 6519 (1995): 227–32; Arto Annila and Erkki Annila, "Why Did Life Emerge?," *International Journal of Astrobiology* 7, no. 3–4 (2008): 293–300; Thomas R. Cech, "The RNA Worlds in Context," *Cold Spring Harbor Perspectives in Biology* 4, no. 7 (2012): a006742; Gerald F. Joyce, "Bit by Bit: The Darwinian Basis of Life," *PLoS Biology* 10, no. 5 (2012): e1001323; Michael P. Robertson and Gerald F. Joyce, "The Origins of the RNA World," *Cold Spring Harbor Perspectives in Biology* 4, no. 5 (2012): a003608; Liudmila S. Yafremava, Monica Wielgos, Suravi Thomas, Arshan Nasir, Minglei Wang, Jay E. Mittenthal, and Gustavo Caetano-Anollés, "A General Framework of Persistence Strategies for Biological Systems Helps Explain Domains of Life," *Frontiers in Genetics* 4 (2013): 16; Robert Pascal, Addy Pross, and John D. Sutherland, "Towards an Evolutionary Theory of the Origin of Life Based on Kinetics and Thermodynamics," *Open Biology* 3, no. 11 (2013): 130156; Arto Annila and Keith Baverstock, "Genes Without Prominence: A Reappraisal of the Foundations of Biology," *Journal of the Royal Society Interface* 11, no. 94 (2014): 20131017; Keith Baverstock and Mauno Rönkkö, "The Evolutionary Origin of Form and Function," *Journal of Physiology* 592, no. 11 (2014): 2261–65; Kepa Ruiz-Mirazo, Carlos Briones, and Andrés de la Escosura,

"Prebiotic Systems Chemistry: New Perspectives for the Origins of Life," *Chemical Reviews* 114, no. 1 (2014): 285–366; Paul G. Higgs and Niles Lehman, "The RNA World: Molecular Cooperation at the Origins of Life," *Nature Reviews Genetics* 16, no. 1 (2015): 7–17; Stuart Kauffman, "What Is Life?," *Israel Journal of Chemistry* 55, no. 8 (2015): 875–79; Abe Pressman, Celia Blanco, and Irene A. Chen, "The RNA World as a Model System to Study the Origin of Life," *Current Biology* 25, no. 19 (2015): R953–R963; Jan Spitzer, Gary J. Pielak, and Bert Poolman, "Emergence of Life: Physical Chemistry Changes the Paradigm," *Biology Direct* 10, no. 33 (2015); Arto Annila and Keith Baverstock, "Discourse on Order vs. Disorder," *Communicative and Integrative Biology* 9, no. 4 (2016): e1187348; Lucas John Mix, "Defending Definitions of Life," *Astrobiology* 15, no. 1 (2015): 15–19; Robert A. Foley, Lawrence Martin, Marta Mirazón Lahr, and Chris Stringer, "Major Transitions in Human Evolution," *Philosophical Transactions of the Royal Society B* 371, no. 1698 (2016): 20150229; Humberto R. Maturana and Francisco J. Varela, "Autopoiesis: The Organization of Living," in *Autopoiesis and Cognition*, ed. Humberto R. Maturana and Francisco J. Varela (Dordrecht: Reidel, 1980), 73–155.

12. Erwin Schrödinger, *What Is Life?* (New York: Macmillan, 1944).

13. Daniel G. Gibson, John I. Glass, Carole Lartigue, Vladimir N. Noskov, Ray-Yuan Chuang, Mikkel A. Algire, Gwynedd A. Benders et al., "Creation of a Bacterial Cell Controlled by a Chemically Synthesized Genome," *Science* 329, no. 5987 (2010): 52–56.

3 VARIETIES OF HOMEOSTASIS

1. Paul Butke and Scott C. Sheridan, "An Analysis of the Relationship Between Weather and Aggressive Crime in Cleveland, Ohio," *Weather, Climate, and Society* 2, no. 2 (2010): 127–39.

2. Joshua S. Graff Zivin, Solomon M. Hsiang, and Matthew J. Neidell, "Temperature and Human Capital in the Short- and Long-Run," *National Bureau of Economic Research* (2015): w21157.

3. Maya E. Kotas and Ruslan Medzhitov, "Homeostasis, Inflammation, and Disease Susceptibility," *Cell* 160, no. 5 (2015): 816–27.

4. Antonio Damasio and Hanna Damasio, "Exploring the Concept of Homeostasis and Considering Its Implications for Economics," *Journal of Economic Behavior & Organization* 2016: 125, 126–29, on which this chapter is partially based; Antonio Damasio, *Self Comes to Mind: Constructing the Conscious Brain* (New York: Pantheon, 2010); Damasio and Carvalho, "Nature of Feelings"; Kent C. Berridge and Morten L. Kringelbach, "Pleasure Systems in the Brain," *Neuron* 86, no. 3 (2015): 646–64.

5. For a brief and intelligent synthesis of this research, see Michael Pollan, "The Intelligent Plant," *New Yorker,* Dec. 23 and 30, 2013; Anthony J. Trewavas, "Aspects of Plant Intelligence," *Annals of Botany* 92, no. 1 (2003): 1–20; Anthony J. Trewavas, "What Is Plant Behaviour?," *Plant, Cell, and Environment* 32, no. 6 (2009): 606–16.

6. John S. Torday, "A Central Theory of Biology," *Medical Hypotheses* 85, no. 1 (2015): 49–57.

7. Claude Bernard, *Leçons sur les phénomènes de la vie communs aux animaux et aux végétaux* (Paris: Librarie J. B. Baillière et Fils, 1879). Reprints from the Collection of the University of Michigan Library.

8. Walter B. Cannon, "Organization for Physiological Homeostasis," *Physiological Reviews* 9, no. 3 (1929): 399–431; Walter B. Cannon, *The Wisdom of the Body* (New York: Norton, 1932); Curt P. Richter, "Total Self-Regulatory Functions in Animals and Human Beings," *Harvey Lecture Series* 38, no. 63 (1943): 1942–43.

9. Bruce S. McEwen, "Stress, Adaptation, and Disease: Allostasis and Allostatic Load," *Annals of the New York Academy of Sciences* 840, no. 1 (1998): 33–44.

10. Trevor A. Day, "Defining Stress as a Prelude to Mapping Its Neurocircuitry: No Help from Allostasis," *Progress in Neuro-psychopharmacology and Biological Psychiatry* 29, no. 8 (2005): 1195–1200.

11. David Lloyd, Miguel A. Aon, and Sonia Cortassa, "Why Homeodynamics, Not Homeostasis?," *Scientific World Journal* 1 (2001): 133–45.

4 FROM SINGLE CELLS TO NERVOUS SYSTEMS AND MINDS

1. Margaret McFall-Ngai, "The Importance of Microbes in Animal Development: Lessons from the Squid-Vibrio Symbiosis," *Annual Review of Microbiology* 68 (2014): 177–94; Margaret McFall-Ngai, Michael G. Hadfield, Thomas C.G. Bosch, Hannah V. Carey, Tomislav Domazet-Lošo, Angela E. Douglas, Nicole Dubilier et al., "Animals in a Bacterial World, a New Imperative for the Life Sciences," *Proceedings of the National Academy of Sciences* 110, no. 9 (2013): 3229–36.

2. Lynn Margulis, *Symbiotic Planet: A New View of Evolution* (New York: Basic Books, 1998).

3. The time period for the likely emergence of circulatory systems, immune systems, and hormonal systems varies remarkably. Circulatory systems begin as early as 700 million years ago. The gastrovascular cavity of cnidarians (about 740 million years ago) are proto-circulatory systems as noted in Eunji Park, Dae-Sik Hwang, Jae-Seong Lee, Jun-Im Song, Tae-Kun Seo, and Yong-Jin Won, "Estimation of Divergence Times in Cnidarian Evolution Based on Mitochondrial Protein-Coding Genes and the Fossil Record," *Molecular Phylogenetics and Evolution* 62, no. 1 (2012): 329–45.

Open circulatory systems allow blood and lymph fluid to mix freely, and were present in arthropods about 600 million years ago (Gregory D. Edgecombe and David A. Legg, "Origins and Early Evolution of Arthropods," *Palaeontology* 57, no. 3 [2014]: 457–68).

The closed circulatory system of vertebrates is characterized by the presence of a cellular barrier—the endothelium—separating tissues from circulating blood. The endothelium evolved in ancestral vertebrates, some 510–540 million years ago, and optimized flow dynamics, barrier function, local immunity, and coagulation (R. Monahan-Earley, A. M. Dvorak, and W. C. Aird, "Evolutionary Origins of the Blood Vascular System and Endothelium," *Journal of Thrombosis and Haemostasis* 11, no. S1 [2013]: 46–66).The innate immune system appears to have begun in cnidarians, during the pre-Cambrian period (Thomas C. G. Bosch, Rene Augustin, Friederike Anton-Erxleben, Sebastian Fraune, Georg Hemmrich, Holger Zill, Philip Rosenstiel et al.,"Uncovering the Evolutionary History of Innate Immunity: The Simple Metazoan Hydra Uses Epithelial Cells for Host Defence," *Developmental and Comparative Immunology* 33, no. 4 [2009]: 559–69).

The adaptive immune system evolved some 450 million years ago in jawed vertebrates (Martin F. Flajnik and Masanori Kasahara, "Origin and Evolution of the Adaptive Immune System: Genetic Events and Selective Pressures," *Nature Reviews Genetics* 11, no. 1 [2010]: 47–59).

Hormonal regulation has far earlier origins, as expected, and can be traced back to unicellular organisms. Bacterial cells "communicate" with hormone-like molecules called autoinducers, which coordinate gene expression (Vanessa Sperandio, Alfredo G. Torres, Bruce Jarvis, James P. Nataro, and James B. Kaper. "Bacteria-Host Communication"). Additionally, insulin-like molecules are found in unicellular organisms (Derek Le Roith, Joseph Shiloach, Jesse Roth, and Maxine A. Lesniak, "Evolutionary Origins of Vertebrate Hormones:

Substances Similar to Mammalian Insulins Are Native to Unicellular Eukaryotes," *Proceedings of the National Academy of Sciences 77*, no. 10 [1980]: 6184–88).

4. For further reading on the operation of neurons, see Eric Kandel, James H. Schwartz, Thomas M. Jessell, Steven A. Siegelbaum, and A. J. Hudspeth, *Principles of Neural Science*, 5th ed. (New York: McGraw-Hill, 2013).

5. František Baluška and Stefano Mancuso, "Deep Evolutionary Origins of Neurobiology: Turning the Essence of 'Neural' Upside-Down," *Communicative and Integrative Biology 2*, no. 1 (2009): 60–65.

6. Damasio and Carvalho, "Nature of Feelings."

7. Anil K. Seth, "Interoceptive Inference, Emotion, and the Embodied Self," *Trends in Cognitive Sciences 17*, no. 11 (2013): 565–73.

8. Andreas Hejnol and Fabian Rentzsch, "Neural Nets," *Current Biology 25*, no. 18 (2015): R782–R786.

9. Detlev Arendt, Maria Antonietta Tosches, and Heather Marlow, "From Nerve Net to Nerve Ring, Nerve Cord, and Brain—Evolution of the Nervous System," *Nature Reviews Neuroscience 17*, no. 1 (2016): 61–72. As will become clear, I am contrasting "intelligence," of which single-celled organisms are abundantly capable, with "mind, consciousness, and feeling," which in my view require nervous systems.

10. For details of neuroanatomy, see Larry W. Swanson, *Brain Architecture: Understanding the Basic Plan* (Oxford: Oxford University Press, 2012); Hanna Damasio, *Human Brain Anatomy in Computerized Images*, 2nd ed. (New York: Oxford University Press, 2005); Kandel et al., *Principles of Neural Science*.

11. We owe this core belief to Warren McCulloch, one of the pioneers of modern neuroscience and one of the founders of computational neuroscience. Had he been with us today he would have been a

vehement critic of his early formulations. Warren S. McCulloch and Walter Pitts, "A Logical Calculus of the Ideas Immanent in Nervous Activity," *Bulletin of Mathematical Biophysics* 5, no. 4 (1943): 115–33; Warren S. McCulloch, *Embodiments of Mind* (Cambridge, Mass.: MIT Press, 1965).

12. Neurons can communicate with other neurons not only by means of synapse but also by "sideways communication mediated by extracellular current flow." The phenomenon is known as ephapsis (see Damasio and Carvalho, "Nature of Feelings," for a hypothesis related to this feature).

5 THE ORIGIN OF MINDS

1. There is abundant evidence to support this idea. For a comprehensive review, see František Baluška and Michael Levin, "On Having No Head: Cognition Throughout Biological Systems," *Frontiers in Psychology* 7 (2016).

2. Sensing and responding relative to the outside are greatly reduced and virtually abolished during deep sleep and deep anesthesia. The interior continues to be sensed and responded to in varied degrees so as to maintain homeostasis. Of note, anesthesia is usually conceived as the negation of consciousness, but that is hardly the case. František Baluška et al., "Understanding of Anesthesia—Why Consciousness Is Essential for Life and Not Based on Genes," *Communicative and Integrative Biology* 9, no. 6 (2016): e1238118.

Apparently, *all* living creatures can be anesthetized, and that includes plants. Anesthesia suspends the processes of sensing and responding. I believe that in complex creatures such as humans anesthesia suspends feelings and consciousness because feelings and consciousness depend on the general machinery of sensing and responding. But feelings and consciousness also depend on other processes; they are not confined to sensing and responding. It is thus *not* possible to conclude that bacteria have feelings and consciousness on the basis of their response to anesthetics. The normal complex behaviors of bacteria do not require feeling or consciousness in the way such phenomena are commonly defined, as I discuss in the chapters ahead.

3. The findings of Tootell and colleagues were illuminating in this regard. Roger B. H. Tootell, Eugene Switkes, Martin S. Silverman, and Susan L. Hamilton, "Functional Anatomy of Macaque Striate Cortex. II. Retinotopic Organization," *Journal of Neuroscience* 8 (1983): 1531–68. See also David Hubel and Torsten Wiesel, *Brain and Visual Perception* (New York: Oxford University Press, 2004); Stephen M. Kosslyn, *Image and Mind* (Cambridge, Mass.: Harvard University Press, 1980); Stephen M. Kosslyn, Giorgio Ganis, and William L. Thompson, "Neural Foundations of Imagery," *Nature Reviews Neuroscience* 2 (2001): 635– 42; Stephen M. Kosslyn, William L. Thompson, Irene J. Kim, and Nathaniel M. Alpert, "Topographical Representations of Mental Images in Primary Visual Cortex," *Nature* 378 (1995): 496–98; Scott D. Slotnick, William L. Thompson, and Stephen M. Kosslyn, "Visual Mental Imagery Induces Retinotopically Organized Activation of Early Visual Areas," *Cerebral Cortex* 15 (2005): 1570–83; Stephen M. Kosslyn, Alvaro Pascual-Leone, Olivier Felician, Susana Camposano, et al. "The Role of Area 17 in Visual Imagery: Convergent Evidence from PET and rTMS," *Science* 284 (1999): 167– 70; Lawrence W. Barsalou, "Grounded Cognition," *Annual Review of Psychology* 59 (2008): 617– 45; W. Kyle Simmons and Lawrence W. Barsalou, "The Similarity-in-Topography Principle: Reconciling Theories of Conceptual Deficits," *Cognitive Neuropsychology* 20 (2003): 451–86; Martin Lotze and Ulrike Halsband, "Motor Imagery," *Journal of Physiology, Paris* 99 (2006): 386–95; Gerald Edelman, *Neural Darwinism: The Theory of Neuronal Group Selection* (New York: Basic Books, 1987), provides a useful discussion of neural maps and insists on the notion of value applied to the selection of maps; Kathleen M.

O'Craven and Nancy Kanwisher, "Mental Imagery of Faces and Places Activates Corresponding Stimulus-Specific Brain Regions," *Journal of Cognitive Neuroscience* 12 (2000): 1013–23; Martha J. Farah, "Is Visual Imagery Really Visual? Overlooked Evidence from Neuropsychology," *Psychological Review* 95 (1988): 307–17; *Principles of Neural Science: Fifth Edition*, edited by Eric Kandel, James H. Schwartz, Thomas M. Jessell, Steven A. Siegelbaum, and A. J. Hudspeth (New York: McGraw-Hill, 2013).

4. Hejnol and Rentzsch, "Neural Nets."

5. Inge Depoortere, "Taste Receptors of the Gut: Emerging Roles in Health and Disease," *Gut* 63, no. 1 (2014): 179–90. To simplify matters, I left out the vestibular sense, which informs us about the body's position in three-dimensional space. The vestibular sense is closely related to hearing, both anatomically and functionally. The sensors are located in the inner ear and thus in the head. Our sense of balance relies on the vestibular system.

6. Signals from each sense are first processed within specialized "early" cortical regions—for example, visual, auditory, somatosensory—but those signals or related signals are subsequently integrated, as needed, in association cortices of the temporal, parietal, and even frontal regions. Each of those regions is interconnected via bidirectional pathways. The processing is further assisted by support networks such as the default mode network and by normal modulation signals hailing from brain-stem nuclei and basal forebrain nuclei. Kingson Man, Antonio Damasio, Kaspar Meyer, & Jonas T. Kaplan, "Convergent and Invariant Object Representations for Sight, Sound, and Touch," *Human Brain Mapping* 36, no. 9 (2015): 3629–40, doi:10.1002/hbm.22867; Kingson Man, Jonas T. Kaplan, Hanna Damasio, and Antonio Damasio, "Neural Convergence and Divergence in the Mammalian Cerebral Cortex: From Experimental Neuroanatomy to Functional Neuroimaging," *Journal of Comparative Neurology* 521, no. 18 (2013): 4097–111, doi:10.1002/cne.23408; Kingson Man, Jonas

T. Kaplan, Antonio Damasio, and Kaspar Meyer, "Sight and Sound Converge to Form Modality-Invariant Representations in Temporo-parietal Cortex," *Journal of Neuroscience* 32, no. 47 (2012): 16629–36, doi:10.1523/JNEUROSCI.2342-12.2012. For background on a neural architecture capable of supporting such processes, see Antonio Damasio et al., "Neural Regionalization of Knowledge Access: Preliminary Evidence," *Symposia on Quantitative Biology* 55 (1990): 1039–47; Antonio Damasio, "Time-Locked Multiregional Retroactivation: A Systems-Level Proposal for the Neural Substrates of Recall and Recognition," *Cognition* 33 (1989): 25–62; Antonio Damasio, Daniel Tranel, and Hanna Damasio, "Face Agnosia and the Neural Substrates of Memory," *Annual Review of Neuroscience* 13 (1990): 89–109. See also Kaspar Meyer and Antonio Damasio, "Convergence and Divergence in a Neural Architecture for Recognition and Memory," *Trends in Neurosciences* 32, no. 7 (2009): 376–82. The discoveries of place cells in hippocampus (by J. O'Keefe) and of grid cells in entorhinal cortex (by M. H. and E. Moser) have expanded our understanding of these systems.

6 EXPANDING MINDS

1. Fernando Pessoa, *The Book of Disquiet* (New York: Penguin Books, 2001).

2. Daydreaming about his own, elusive success, the character of Oscar Levant, a composer, imagines himself in a concert hall playing the piano for a public made up of several Oscar Levants, enthusiastically applauding, of course. Eventually, he plays other instruments and conducts, too.

3. The simplification of the account of periphery/brain relationships is one of the main problems facing any attempt to understand mental processes in biological terms. The actual process violates the traditional conception of the brain as a separate organ that receives computerlike signals and responds as needed. The reality is that the

signals are never purely neural to begin with and gradually change along the way to the central nervous system. Moreover, the nervous system can respond to the incoming signals at varied levels and thus alter the original conditions that gave rise to the signaling.

4. The investigation of the neural basis of concept and language processing has been one of the main areas of research in cognition neuroscience. Our group has contributed to the field, and these references point to some of the contributions we have made over the years: Antonio Damasio and Patricia Kuhl, "Language," in Kandel et al., *Principles of Neural Science;* Hanna Damasio, Daniel Tranel, Thomas J. Grabowski, Ralph Adolphs, and Antonio Damasio, "Neural Systems Behind Word and Concept Retrieval," *Cognition* 92, no. 1 (2004): 179–229; Antonio Damasio and Daniel Tranel, "Nouns and Verbs Are Retrieved with Differently Distributed Neural Systems," *Proceedings of the National Academy of Sciences* 90, no. 11 (1993): 4957–60; Antonio Damasio, "Concepts in the Brain," *Mind and Language* 4, nos. 1–2 (1989): 24–28, doi:10.1111/j.1468-0017.tb00236.x; Antonio Damasio and Hanna Damasio, "Brain and Language," *Scientific American* 267 (1992): 89–95.

5. The neural correlates of the process of constructing narratives can now be investigated in the laboratory. See as an example Jonas Kaplan, Sarah I. Gimbel, Morteza Dehghani, Mary Helen Immordino-Yang, Kenji Sagae, Jennifer D. Wong, Christine Tipper, Hanna Damasio, Andrew S. Gordon, and Antonio Damasio, "Processing Narratives Concerning Protected Values: A Cross-Cultural Investigation of Neural Correlates," *Cerebral Cortex* (2016): 1–11, doi:10.1093/cercor /bhv325.

6. "Default mode network" refers to a set of bilateral cortical regions that become especially active in certain behavioral and mental conditions, such as resting and mind wandering, and may become less active when the mind is focused on a particular content. Or not, because in

some conditions of attentive processing the network actually becomes *more* active. The nodes in the network correspond to regions of high convergence and divergence of cortical connections within what is traditionally known as association cortices. The network probably plays a role in the organization of mental contents in the process of memory search and narrative compositions. Many of the features of this network (and of other related networks) are puzzling. Marcus Raichle's careful observations led to its discovery. Marcus E. Raichle, "The Brain's Default Mode Network," *Annual Review of Neuroscience* 38 (2015): 433–47.

7. Meyer and Damasio, "Convergence and Divergence in a Neural Architecture for Recognition and Memory" and related articles on convergence—divergence framework.

8. The philospher Avishai Margalit has made an important contribution to the study of these issues, see *The Ethics of Memory* (Cambridge, Mass.: Harvard University Press, 2002).

7 AFFECT

1. See *Descartes' Error* for an early description of "as-if body loop." Lisa Feldman Barrett's description of feelings captures my idea of intellectualized feelings. It calls attention to an elaboration of the basic feeling process that relies on memory and reasoning. Lisa Feldman Barrett, Batja Mesquita, Kevin N. Ochsner, and James J. Gross, "The Experience of Emotion," *Annual Review of Psychology* 58 (2007): 373.

2. I am making a principled distinction between the mental contents that belong to the basic feeling process—for example, valence—and the mental contents that belong to the intellectualization of the process: memories, reasonings, descriptions. I am letting Caesar have what belongs to Caesar and no more.

3. Lauri Nummenmaa, Enrico Glerean, Riitta Hari, and Jari K. Hietanen, "Bodily Maps of Emotions," *Proceedings of the National Academy of Sciences* 111, no. 2 (2014): 646–51.

4. William Wordsworth, "Lines Composed a Few Miles Above Tintern Abbey, on Revisiting the Banks of the Wye During a Tour, July 13, 1798," in *Lyrical Ballads* (Monmouthshire, U.K.: Old Stile Press, 2002), 111–17.

5. Personal communication from Mary Helen Immordino-Yang.

6. Rewarding physiological conditions are associated with the release of endogenous endorphin molecules, which are agonists for the mu opioid receptor (MOR). MORs are best known in the context of analgesia and drug addiction, but more recently they have been credited with mediating the pleasurable quality of rewarding experiences. Morten L. Kringelbach and Kent C. Berridge, "Motivation and Pleasure in the Brain," in *The Psychology of Desire,* ed. Wilhelm Hofmann and Loran F. Nordgren (New York: Guilford Press, 2015), 129–45.

7. By definition, stress is metabolically intensive, and recent studies have demonstrated that while acute stress can increase the strength of an immune response, chronic stress has the opposite effect, inhibiting an organism's ability to fight off immune challenges. Engaging immune responses mobilizes the cellular factories that produce immune cells. The process is metabolically expensive, and mounting an effective immune response sometimes requires more resources than an organism can easily spare, especially if they are already in a stressed state. When this occurs, an organism's well-being deteriorates, and as other homeostatic budgets are slashed to support the defense effort, exhaustion and lethargy set in, further reducing the chances of full recovery. In this frame, it is apparent that a non-stressed organism has the best chances of mounting an effective immune response, and therefore the best chances of sustaining a state of flourishing.

See Terry L. Derting and Stephen Compton, "Immune Response, Not Immune Maintenance, Is Energetically Costly in Wild White-Footed Mice (*Peromyscus leucopus*)," *Physiological and Biochemical Zoology* 76, no. 5 (2003): 744–52; Firdaus S. Dhabhar and Bruce S. McEwen, "Acute Stress Enhances While Chronic Stress Suppresses Cell-Mediated Immunity in Vivo: A Potential Role for Leukocyte Trafficking," *Brain, Behavior, and Immunity* 11, no. 4 (1997): 286–306; Suzanne C. Segerstrom and Gregory E. Miller, "Psychological Stress and the Human Immune System: A Meta-analytic Study of 30 Years of Inquiry," *Psychological Bulletin* 130, no. 4 (2004): 601.

Stress activates the hypothalamic-pituitary axis, and induces corticotropin-releasing hormone (CRH), which binds to the CRH1 receptor, and prompts the release of dynorphin, a different class of endogenous opioid peptide. Dynorphin is a kappa opioid receptor (KOR) agonist, and while MORs have been associated with the pleasant quality of rewarding experiences, KOR activity in the basolateral amygdala has been credited with mediating the aversive quality of unpleasant experience. See Benjamin B. Land et al., "The Dysphoric Component of Stress Is Encoded by Activation of the Dynorphin K-Opioid System," *Journal of Neuroscience* 28, no. 2 (2008): 407–14; Michael R. Bruchas, Benjamin B. Land, Julia C. Lemos, and Charles Chavkin, "CRF1-R Activation of the Dynorphin/Kappa Opioid System in the Mouse Basolateral Amygdala Mediates Anxiety-Like Behavior," *PLoS One* 4, no. 12 (2009): e8528.

8. Jaak Panksepp has made pioneering contributions to the understanding of the role of brain stem and basal forebrain structure in affect. See Panksepp, *Affective Neuroscience*; other relevant work includes Antonio Damasio, Thomas J. Grabowski, Antoine Bechara, Hanna Damasio, Laura L.B. Ponto, Josef Parvizi, and Richard Hichwa, "Subcortical and Cortical Brain Activity During the Feeling of Self-Generated Emotions," *Nature Neuroscience* 3, no. 10 (2000): 1049–56, doi:10.1038/79871; Antonio Damasio and Joseph LeDoux, "Emotion,"

in Kandel et al., *Principles of Neural Science*. See Berridge and Kringel-bach, *Pleasures of the Brain* (Oxford: Oxford University Press, 2009); Damasio and Carvalho, "Nature of Feelings"; Josef Parvizi and Anto-nio Damasio, "Consciousness and the Brainstem," *Cognition* 79, no. 1 (2001): 135–60, doi:10.1016/S0010-0277(00)00127-X. For a recent review, see Anand Venkatraman, Brian L. Edlow, and Mary Helen Immordino-Yang, "The Brainstem in Emotion: A Review," *Frontiers in Neuroanatomy* 11, no. 15 (2017): 1–12; Jaak Panksepp, "The Basic Emotional Circuits of Mammalian Brains: Do Animals Have Affec-tive Lives?," *Neuroscience and Biobehavioral Reviews* 35, no. 9 (2011): 1791–804; Antonio Alcaro and Jaak Panksepp, "The SEEKING Mind: Primal Neuro-affective Substrates for Appetitive Incentive States and Their Pathological Dynamics in Addictions and Depression," *Neurosci-ence and Biobehavioral Reviews* 35, no. 9 (2011): 1805–20; Stephen M. Siviy and Jaak Panksepp, "In Search of the Neurobiological Sub-strates for Social Playfulness in Mammalian Brains," *Neuroscience and Biobehavioral Reviews* 35, no. 9 (2011): 1821–30; Jaak Panksepp, "Cross-Species Affective Neuroscience Decoding of the Primal Affec-tive Experiences of Humans and Related Animals," *PLoS One* 6, no. 9 (2011): e21236.

9. When you hear a scream and end up reacting to it by feeling some variation of fear, the mechanism behind that emotional feeling is grounded on an emotive response triggered by the acoustical char-acteristics of the scream; the high pitch of the sound may contribute to the response, but as it now seems to be the case, the *roughness* of the sound appears to be the critical element. The circumstances in which you hear the scream are also relevant. If I hear Janet Leigh screaming in Orson Welles's *Touch of Evil* (or in Hitchcock's *Psycho*), which I have seen many times, it is a scream I fully expect; the nega-tive emotive response still occurs but is muted; I may even override the negative feeling with a positive one as I watch how Welles edited the scene. But if I hear a similar scream when I am alone at night in the

alley where I had to park my car, it will be a different story. I will be frightened. I will have some variation of the "conventional" emotion fear and of the consequent feeling of fear. The inevitable consequence of deploying an emotive program is a modification of some aspects of the ongoing homeostatic state. The mental representation—the imaging—of this process of modification and its durable or fleeting culmination are emotional feelings, the standard variety of provoked feeling. Luc H. Arnal, Adeen Flinker, Andreas Kleinschmidt, Anne-Lise Giraud, and David Poeppel, "Human Screams Occupy a Privileged Niche in the Communication Soundscape," *Current Biology* 25, no. 15 (2015): 2051–56; Ralph Adolphs, Hanna Damasio, Daniel Tranel, Greg Cooper, and Antonio Damasio, "A Role for Somatosensory Cortices in the Visual Recognition of Emotion as Revealed by Three-Dimensional Lesion Mapping," *Journal of Neuroscience* 20, no. 7 (2000): 2683–90.

10. The "desire" for social relationships is, unsurprisingly, both ancient and homeostatically motivated. Single-celled organisms exhibit precursors of these phenomena, and we can find other examples in birds and mammals.

In the wild, increased parasite transmission and resource competition among social animals can reduce reproductive success and longevity. This can be offset by social grooming, an adaptive behavior that not only minimizes parasite load but forges social bonds and alliances between grooming partners. Among certain primates, social grooming is the centerpiece of complex systems of social hierarchy, reciprocity, and resource/service exchanges. Social relationships formed around grooming partnerships are vital for individual health and well-being and support group cohesion. See Cyril C. Greuter, Annie Bissonnette, Karin Isler, and Carel P. van Schaik, "Grooming and Group Cohesion in Primates: Implications for the Evolution of Language," *Evolution and Human Behavior* 34, no. 1 (2013): 61–68; Karen McComb and Stuart Semple, "Coevolution of Vocal Communication and Sociality in Primates," *Biology Letters* 1, no. 4 (2005): 381–85; Max Henning, Glenn R. Fox, Jonas Kaplan, Hanna Damasio, and Antonio Damasio, "A Role for mu-Opioids in Mediating the Positive Effects of Gratitude," in *Focused Review: Frontiers in Psychology* (forthcoming).

11. Social play behavior is mediated by subcortical brain circuitry. Research has shown that rough-and-tumble play among juvenile animals is critical for learning what constitutes acceptable social behavior. Domestic kittens deprived of social play become aggressive adult cats. Additionally, social play behavior appears to be modulated by the opioidergic mechanisms, with mu and kappa opioid receptor activation exerting facilitative or inhibitory effects. These opioid mechanisms are more generally associated with homeostatic drives and affective valence; their involvement in sociality suggests that pro-social behavior is homeostatically motivated. Siviy and Panksepp, "In Search of the Neurobiological Substrates for Social Playfulness in Mammalian Brains"; Panksepp, "Cross-Species Affective Neuroscience Decoding of the Primal Affective Experiences of Humans and Related Animals"; Gary W. Guyot, Thomas L. Bennett, and Henry A. Cross, "The Effects of Social Isolation on the Behavior of Juvenile Domestic Cats," *Developmental Psychobiology* 13, no. 3 (1980): 317–29; Louk J. M. J. Vanderschuren, Raymond J. M. Niesink, Berry M. Spruijt, and Jan M. Van Ree, "μ-and κ-Opioid Receptor-Mediated Opioid Effects on Social Play in Juvenile Rats," *European Journal of Pharmacology* 276, no. 3 (1995): 257–66; Hugo A. Tejeda, Danielle S. Counotte, Eric Oh, Sammanda Ramamoorthy, Kristin N. Schultz-Kuszak, Cristina M. Bäckman, Vladmir Chefer, Patricio O'Donnell, and Toni S. Shippenberg, "Prefrontal Cortical Kappa-Opioid Receptor Modulation of Local Neurotransmission and Conditioned Place Aversion," *Neuropsychopharmacology* 38, no. 9 (2013): 1770–79; Stephen W. Porges, *The Polyvagal Theory* (New York and London: W. W. Norton, 2011).

In recent studies of species endowed with neural systems necessary for image making, positive and negative valence have been consistently correlated with mu and kappa opioid receptors, respectively. The quartet of opioid receptors—delta, mu, kappa, and NOP—in the human body has been conserved since jawed vertebrates first made their appearance after the Cambrian explosion, about 450 million years ago, and it is intriguing to consider the possibility that valence, and even feeling, may be far more pervasive in the animal kingdom than has been conventionally thought. Susanne Dreborg, Görel Sundström, Tomas A. Larsson, and Dan Larhammar, "Evolution of Vertebrate Opioid Receptors," *Proceedings of the National Academy of Sciences* 105, no. 40 (2008): 15487–92.

8 THE CONSTRUCTION OF FEELINGS

1. Pierre Beaulieu et al., *Pharmacology of Pain* (Philadelphia: Lippincott Williams & Wilkins, 2015).

2. George B. Stefano, Beatrice Salzet, and Gregory L. Fricchione, "Enkelytin and Opioid Peptide Association in Invertebrates and Vertebrates: Immune Activation and Pain," *Immunology Today* 19, no. 6 (1998): 265–68; Michel Salzet and Aurélie Tasiemski, "Involvement of Pro-enkephalin-derived Peptides in Immunity," *Developmental and Comparative Immunology* 25, no. 3 (2001): 177–85; Halina Machelska and Christoph Stein, "Leukocyte-Derived Opioid Peptides and Inhibition of Pain," *Journal of Neuroimmune Pharmacology* 1, no. 1 (2006): 90–97; Simona Farina, Michele Tinazzi, Domenica Le Pera, and Massimiliano Valeriani, "Pain-Related Modulation of the Human Motor Cortex," *Neurological Research* 25, no. 2 (2003): 130–42; Stephen B. McMahon, Federica La Russa, and David L. H. Bennett, "Crosstalk Between the Nociceptive and Immune Systems in Host Defense and Disease," *Nature Reviews Neuroscience* 16, no. 7 (2015): 389–402.

3. Brunet and Arendt, "From Damage Response to Action Potentials"; Hoffman et al., "Aminoglycoside Antibiotics Induce Bacterial

Biofilm Formation"; Naviaux, "Metabolic Features of the Cell Danger Response"; Icard-Arcizet et al., "Cell Stiffening in Response to External Stress Is Correlated to Actin Recruitment"; Kearns, "Field Guide to Bacterial Swarming Motility"; Erill, Campoy, and Barbé, "Aeons of Distress."

Transient receptor potential (TRP) ion channels serve as the sensors in single-celled organisms and are conserved throughout phylogeny. In invertebrates, for example, these sensors can detect noxious environmental conditions such as intense heat, and thus are critical for navigating safely. The combination of noxious detection devices with nervous systems has eventually led to an entire class of sensory neurons called nociceptors.

Nociceptors are distributed throughout body tissue and are equipped with high-threshold TRP ion channels that respond to noxious intensities of otherwise harmless sensations. Nociceptors are also equipped with toll-like receptors (TLR), the sentinels of the immune system that are distributed throughout the body. Activation of TLRs induces an immune response, and when nociceptor TLRs are activated, they induce a potent, localized inflammatory response, and they sensitize local nociceptive TRP channels, contributing to the increased pain sensitivity associated with injury or infection. Pain, in turn, inhibits motor cortex and has even been shown to inhibit the initiation of motion itself by activating antagonistic muscle groups. In the case of injury, this could prevent additional damage.

Nociceptive sensory afferents address pain and damage while non-nociceptive sensory afferents gather other relevant information about the conditions inside and outside the organism, resulting in images that are processed simultaneously. Nervous systems allow for the precise localization of sensory stimulation and for the coordination of complex and diverse physiological processes that integrate all major life-regulatory systems in the homeostatic effort. Giorgio Santoni, Claudio Cardinali, Maria Beatrice Morelli, Matteo Santoni, Massimo

Nabissi, and Consuelo Amantini, "Danger- and Pathogen-Associated Molecular Patterns Recognition by Pattern-Recognition Receptors and Ion Channels of the Transient Receptor Potential Family Triggers the Inflammasome Activation in Immune Cells and Sensory Neurons," *Journal of Neuroinflammation* 12, no. 1 (2015): 21; McMahon, La Russa, and Bennett, "Crosstalk Between the Nociceptive and Immune Systems in Host Defense and Disease"; Ardem Patapoutian, Simon Tate, and Clifford J. Woolf, "Transient Receptor Potential Channels: Targeting Pain at the Source," *Nature Reviews Drug Discovery* 8, no. 1 (2009): 55–68; Takaaki Sokabe and Makoto Tominaga, "A Temperature-Sensitive TRP Ion Channel, Painless, Functions as a Noxious Heat Sensor in Fruit Flies," *Communicative and Integrative Biology* 2, no. 2 (2009): 170–73; Farina et al., "Pain-Related Modulation of the Human Motor Cortex."

4. Santoni et al., "Danger- and Pathogen-Associated Molecular Patterns Recognition by Pattern-Recognition Receptors and Ion Channels of the Transient Receptor Potential Family Triggers the Inflammasome Activation in Immune Cells and Sensory Neurons"; Sokabe and Tominaga, "Temperature-Sensitive TRP Ion Channel, Painless, Functions as a Noxious Heat Sensor in Fruit Flies."

5. Colin Klein and Andrew B. Barron, "Insects Have the Capacity for Subjective Experience," *Animal Sentience* 1, no. 9 (2016): 1.

Although nerve nets in hydras were probably not capable of producing images or even representations, an intermediate step was emerging. Toll-like receptors (TLRs), the internal receptors whose activation signals the presence of invading pathogens or tissue damage from heat shock or other noxious conditions, are found in hydras and thus antecede nervous-system-dependent mapping. The specific sensitivity of TLRs to damage- or pathogen-associated molecular patterns allows TLR activation to provoke specific emotive and innate immune response. This specificity in detection/response is a step up from the generalized sensations facilitated by transient receptor potential ion channels that are present in single-celled organisms. Sören Franzenburg, Sebastian Fraune, Sven Künzel, John F. Baines, Tomislav Domazet-Lošo, and Thomas C. G. Bosch, "MyD88-Deficient Hydra Reveal an Ancient Function of TLR Signaling in Sensing Bacterial Colonizers," *Proceedings of the National Academy of Sciences* 109, no. 47 (2012): 19374–79; Bosch et al., "Uncovering the Evolutionary History of Innate Immunity."

6. Feelings can make the difference between life and death. Every living organism must respond to environmental conditions as they are detected, but there are many cases in which the time it takes to identify the homeostatically relevant quality of an environment is a matter of survival. An animal that can predict the presence of predators from familiar environmental cues has better chances of survival, and feelings allow for just that.

Studies of the phenomenon of conditioned place aversion/preference address this issue. An experimental animal is trained to associate neutral environmental cues with a homeostatically relevant stimulus such that the environmental cues themselves begin to induce the response even in the absence of the homeostatically relevant stimulus. It is unlikely that this kind of flexible learning occurs in non-feeling organisms. For it to occur, there must first be an internal representation of the specific environmental cues, as well as a representation of physiological distress, so that the two models can be joined. The next time the environmental cues are detected, they will provoke the associated physiological state.

Feeling capabilities allow an animal to react predictively to perceived conditions of the external environment in a manner that reflects its own past experiences. This projection of subjective homeostatic relevance onto otherwise neutral environmental stimuli allows for significant increases in organism survivability and productivity. See

Cindee F. Robles, Marissa Z. McMackin, Katharine L. Campi, Ian E. Doig, Elizabeth Y. Takahashi, Michael C. Pride, and Brian C. Trainor, "Effects of Kappa Opioid Receptors on Conditioned Place Aversion and Social Interaction in Males and Females," *Behavioural Brain Research* 262 (2014): 84–93; M. T. Bardo, J. K. Rowlett, and M. J. Harris, "Conditioned Place Preference Using Opiate and Stimulant Drugs: A Meta-analysis," *Neuroscience and Biobehavioral Reviews* 19, no. 1 (1995): 39–51.

7. While activation of the *innate immune system* induces a generalized protective response to any form of tissue damage or infection, the *adaptive immune system*—which evolved later, in jawed vertebrates about 450 million years ago—mounts a direct assault targeting a specific pathogen. Once a pathogen has been identified for the first time, specific molecules are produced that are selective for that pathogen. When the pathogen is subsequently detected by these molecules, an army of immune cells is quickly generated that sweeps the body searching for any cells that bear the molecular signature of the invader. These signatures are remembered for the life of the organism, and repeated exposure to pathogens strengthens the adaptive immune responses over time. Martin F. Flajnik and Masanori Kasahara, "Origin and Evolution of the Adaptive Immune System: Genetic Events and Selective Pressures," *Nature Reviews Genetics* 11, no. 1 (2010): 47–59.

8. Klein and Barron, "Insects Have the Capacity for Subjective Experience."

9. Yasuko Hashiguchi, Masao Tasaka, and Miyo T. Morita, "Mechanism of Higher Plant Gravity Sensing," *American Journal of Botany* 100, no. 1 (2013): 91–100; Alberto P. Macho and Cyril Zipfel, "Plant PRRs and the Activation of Innate Immune Signaling," *Molecular Cell* 54, no. 2 (2014): 263–72.

10. My colleague Kingson Man suggested the term "continuity" to denote the conditions under which neural-body interactions take place.

11. Traditional Eastern metaphysical systems of thought contend that while duality is inherent to the normal mode of human perception, the world we perceive—full of discrete and independent objects or phenomena—is a perceptual screen masking a more fundamental, "non-dual" substrate of reality. "Non-duality" describes a world of absolute interdependence, in which mind, body, and all phenomena are inextricable. Although this view is incompatible with to the dominant cultural paradigms of the West, some Western philosophers—Spinoza in particular—arrived at similar conclusions. Parallels between these pillars of traditional Eastern thought and the current natural sciences continue to be uncovered. Consider, for example, the remarkable discoveries in quantum physics suggesting that underneath the discretized, objectified reality we perceive with our senses, there lies a more relational, dynamic interplay of forces that challenges dominant view. David Loy, *Nonduality: A Study in Comparative Philosophy* (Amherst, N.Y.: Humanity Books, 1997); Vlatko Vedral, *Decoding Reality: The Universe as Quantum Information* (New York: Oxford University Press, 2012).

12. Arthur D. Craig, "How Do You Feel? Interoception: The Sense of the Physiological Condition of the Body," *Nature Reviews Neuroscience* 3, no. 8 (2002): 655–66; Arthur D. Craig, "Interoception: The Sense of the Physiological Condition of the Body," *Current Opinion in Neurobiology* 13, no. 4 (2003): 500–505; Arthur D. Craig, "How Do You Feel—Now? The Anterior Insula and Human Awareness," *Nature Reviews Neuroscience* 10, no. 1 (2009); Hugo D. Critchley, Stefan Wiens, Pia Rotshtein, Arne Öhman, and Raymond J. Dolan, "Neural Systems Supporting Interoceptive Awareness," *Nature Neuroscience* 7, no. 2 (2004): 189–95.

13. Alexander J. Shackman, Tim V. Salomons, Heleen A. Slagter, Andrew S. Fox, Jameel J. Winter, and Richard J. Davidson, "The Integration of Negative Affect, Pain, and Cognitive Control in the Cingulate Cortex," *Nature Reviews Neuroscience* 12, no. 3 (2011): 154–67.

14. Jaak Panksepp was the champion of the subcortical nuclei at a time no one paid attention to them. The idea has received ample support, including from our own work: Damasio et al., "Subcortical and Cortical Brain Activity During the Feeling of Self-Generated Emotions." The primate anatomy of the brain stem has been well explained in Parvizi and Damasio, "Consciousness and the Brainstem."

15. The importance of these nuclei can be gleaned by the massive projections they receive concerning homeostatic state changes. Esther-Marije Klop, Leonora J. Mouton, Rogier Hulsebosch, José Boers, and Gert Holstege, "In Cat Four Times as Many Lamina I Neurons Project to the Parabrachial Nuclei and Twice as Many to the Periaqueductal Gray as to the Thalamus," *Neuroscience* 134, no. 1 (2005): 189–97.

16. Michael M. Behbehani, "Functional Characteristics of the Midbrain Periaqueductal Gray," *Progress in Neurobiology* 46, no. 6 (1995): 575–605.

17. Craig, "How Do You Feel?"; Craig, "Interoception"; Craig, "How Do You Feel—Now?"; Critchley et al., "Neural Systems Supporting Interoceptive Awareness"; Richard P. Dum, David J. Levinthal, and Peter L. Strick, "The Spinothalamic System Targets Motor and Sensory Areas in the Cerebral Cortex of Monkeys," *Journal of Neuroscience* 29, no. 45 (2009): 14223–35; Antoine Louveau, Igor Smirnov, Timothy J. Keyes, Jacob D. Eccles, Sherin J. Rouhani, J. David Peske, Noel C. Derecki, "Structural and Functional Features of Central Nervous System Lymphatic Vessels," *Nature* 523, no. 7560 (2015): 337–41.

18. Michael J. McKinley, *The Sensory Circumventricular Organs of the Mammalian Brain: Subfornical Organ, OVLT, and Area Postrema* (New York: Springer, 2003); Robert E. Shapiro and Richard R. Miselis, "The Central Neural Connections of the Area Postrema of the Rat," *Journal of Comparative Neurology* 234, no. 3 (1985): 344–64.

19. Marshall Devor, "Unexplained Peculiarities of the Dorsal Root Ganglion," *Pain* 82 (1999): S27–S35.

20. He-Bin Tang, Yu-Sang Li, Koji Arihiro, and Yoshihiro Nakata, "Activation of the Neurokinin-1 Receptor by Substance P Triggers the Release of Substance P from Cultured Adult Rat Dorsal Root Ganglion Neurons," *Molecular Pain* 3, no. 1 (2007): 42.

21. J. A. Kiernan, "Vascular Permeability in the Peripheral Autonomic and Somatic Nervous Systems: Controversial Aspects and Comparisons with the Blood-Brain Barrier," *Microscopy Research and Technique* 35, no. 2 (1996): 122–36.

22. Malin Björnsdotter, India Morrison, and Håkan Olausson, "Feeling Good: On the Role of C Fiber Mediated Touch in Interoception," *Experimental Brain Research* 207, no. 3–4 (2010): 149–55; A. Harper and S. N. Lawson, "Conduction Velocity Is Related to Morphological Cell Type in Rat Dorsal Root Ganglion Neurones," *Journal of Physiology* 359 (1985): 31.

23. Damasio and Carvalho, "Nature of Feelings"; Ian A. McKenzie, David Ohayon, Huiliang Li, Joana Paes De Faria, Ben Emery, Koujiro Tohyama, and William D. Richardson, "Motor Skill Learning Requires Active Central Myelination," *Science* 346, no. 6207 (2014): 318–22.

24. Ongoing research in our group indicates that non-synaptic transmission in peripheral nervous system ganglia is controlled by a ubiquitous neurotransmitter that also plays a key role in synaptic transmission, as well as in pain, sensory perception, smooth muscle contraction, and a host of other bodily functions. Interestingly, this multifaceted molecule does not affect neurons indiscriminately. It seems to reserve its most dramatic impact for the ancient, unmyelinated C-type neurons that form most of our interoceptive pathways and are likely to play a role in the generation of feelings. See Damasio and Carvalho, "Nature of Feelings"; Björnsdotter, Morrison, and Olausson, "Feeling Good"; Gang Wu, Matthias Ringkamp, Timothy V. Hartke, Beth B. Murinson, James N. Campbell, John W. Griffin, and Richard A. Meyer, "Early Onset of Spontaneous Activity in Uninjured

C-Fiber Nociceptors After Injury to Neighboring Nerve Fibers," *Journal of Neuroscience* 21, no. 8 (2001): RC140; R. Douglas Fields, "White Matter in Learning, Cognition, and Psychiatric Disorders," *Trends in Neurosciences* 31, no. 7 (2008): 361–70; McKenzie et al., "Motor Skill Learning Requires Active Central Myelination"; Julia J. Harris and David Attwell, "The Energetics of CNS White Matter," *Journal of Neuroscience* 32, no. 1 (2012): 356–71; Richard A. Meyer, Srinivasa N. Raja, and James N. Campbell, "Coupling of Action Potential Activity Between Unmyelinated Fibers in the Peripheral Nerve of Monkey," *Science* 227 (1985): 184–88; Hemant Bokil, Nora Laaris, Karen Blinder, Mathew Ennis, and Asaf Keller, "Ephaptic Interactions in the Mammalian Olfactory System," *Journal of Neuroscience* 21 (2001): 1–5; Henry Harland Hoffman and Harold Norman Schnitzlein, "The Numbers of Nerve Fibers in the Vagus Nerve of Man," *Anatomical Record* 139, no. 3 (1961): 429–35; Marshall Devor and Patrick D. Wall, "Cross-Excitation in Dorsal Root Ganglia of Nerve-Injured and Intact Rats," *Journal of Neurophysiology* 64, no. 6 (1990): 1733–46; Eva Sykova, "Glia and Volume Transmission During Physiological and Pathological States," *Journal of Neural Transmission* 112, no. 1 (2005): 137–47.

25. Emeran Mayer, *The Mind-Gut Connection: How the Hidden Conversation Within Our Bodies Impacts Our Mood, Our Choices, and Our Overall Health* (New York: HarperCollins, 2016).

26. Jane A. Foster and Karen-Anne McVey Neufeld, "Gut-Brain Axis: How the Microbiome Influences Anxiety and Depression," *Trends in Neurosciences* 36, no. 5 (2013): 305–12; Mark Lyte and John F. Cryan, eds., *Microbial Endocrinology: The Microbiota-Gut-Brain Axis in Health and Disease* (New York: Springer, 2014); Mayer, *Mind-Gut Connection.*

27. Doe-Young Kim and Michael Camilleri, "Serotonin: A Mediator of the Brain-Gut Connection," *American Journal of Gastroenterology* 95, no. 10 (2000): 2698.

28. Timothy R. Sampson, Justine W. Debelius, Taren Thron, Stefan Janssen, Gauri G. Shastri, Zehra Esra Ilhan, Collin Challis et al., "Gut Microbiota Regulate Motor Deficits and Neuroinflammation in a Model of Parkinson's Disease," *Cell* 167, no. 6 (2016): 1469–80.

29. Sadness can certainly disturb health, but positive states such as gratitude appear to have the opposite effect. Gratitude is induced when we receive meaningful aid or support that is motivated by compassion and is associated with significant positive effects on health and quality of life. Recently, an fMRI study by my colleague Glenn Fox defined the neural correlates of gratitude, revealing that the reported experience of meaningful gratitude is correlated with brain activity in regions conventionally recognized as central to stress regulation, social cognition, and moral reasoning. This finding supports previous research showing that developing gratitude as a mental habit can improve health, which in turn underscores the idea of continuity between the mind and the body. See Glenn R. Fox, Jonas Kaplan, Hanna Damasio, and Antonio Damasio, "Neural Correlates of Gratitude," *Frontiers in Psychology* 6 (2015); Alex M. Wood, Stephen Joseph, and John Maltby, "Gratitude Uniquely Predicts Satisfaction with Life: Incremental Validity Above the Domains and Facets of the Five Factor Model," *Personality and Individual Differences* 45, no. 1 (2008): 49–54; Max Henning, Glenn R. Fox, Jonas Kaplan, Hanna Damasio, and Antonio Damasio, "The Positive Effects of Gratitude Are Mediated by Physiological Mechanisms," *Frontiers in Psychology* (2017).

30. Sarah J. Barber, Philipp C. Opitz, Bruna Martins, Michiko Sakaki, and Mara Mather, "Thinking About a Limited Future Enhances the Positivity of Younger and Older Adults' Recall: Support for Socioemotional Selectivity Theory," *Memory and Cognition* 44, no. 6 (2016): 869–82; Mara Mather, "The Affective Neuroscience of Aging," *Annual Review of Psychology* 67 (2016): 213–38.

31. Daniel Kahneman, "Experienced Utility and Objective Happiness: A Moment-Based Approach," in *Choices, Values, and Frames,*

eds. Daniel Kahneman and Amos Tversky (New York: Russell Sage Foundation, 2000); Daniel Kahneman, "Evaluation by Moments: Past and Future," in ibid.; Bruna Martins, Gal Sheppes, James J. Gross, and Mara Mather, "Age Differences in Emotion Regulation Choice: Older Adults Use Distraction Less Than Younger Adults in High-Intensity Positive Contexts," *Journals of Gerontology Series B: Psychological Sciences and Social Sciences* (2016): gbw028.

9 CONSCIOUSNESS

1. Two brief notes: first, I am using the term "subjectivity" in its cognitive and philosophical sense and not in the popular sense, where "subjective" refers to "personal opinion"; second, I have been working on the problems of consciousness for many years and presented some of my ideas in two books: *The Feeling of What Happens* and *Self Comes to Mind.* Subsequent publications have introduced extensions of those ideas. See Antonio Damasio, Hanna Damasio, and Daniel Tranel, "Persistence of Feelings and Sentience After Bilateral Damage of the Insula," *Cerebral Cortex* 23 (2012): 833–46; Damasio and Carvalho, "Nature of Feelings"; Antonio Damasio and Hanna Damasio, "Pain and Other Feelings in Humans and Animals," *Animal Sentience* 1, no. 3 (2016): 33. My views continued to evolve, influenced by parallel efforts on theoretical and empirical work on disorders of feeling and consciousness, but this is not the place to present the latest developments that will be the object of a separate volume.

2. The designation "Cartesian Theater" comes from Daniel Dennett's spirited discussions of consciousness, which include a clear and welcome dismissal of the "homunculi" myths and a warning about the dangers of infinite regress—the idea that a little person would sit in our brains and survey the mind, followed by the need to postulate

another little person who would survey the former, and so forth ad infinitum.

3. I used to deal with the topic of subjectivity by appealing to the term "self," but now I refrain from using the term to avoid the possible impression, entirely unjustified, that self, from simple to complex levels, is some sort of fixed, well-bounded object or center of control. One should never underestimate the potential for ill-disposed interpretations of the notion of self as a homunculus. The ensuing confusion, even if one mentions not a word about the neuroanatomical correlates of the self phenomena, conjures up the specter of phrenology.

4. A number of colleagues have advanced accounts of mental integration that are generally compatible with mine, most prominently Bernard Baars, Stanislas Dehaene, and Jean-Pierre Changeux. Their ideas are clearly discussed in Stanislas Dehaene, *Consciousness and the Brain: Deciphering How the Brain Codes Our Thoughts* (New York: Viking, 2014).

5. This applies to an elusive brain area known as the claustrum, championed by Francis Crick and Christof Koch, "A Framework for Consciousness," *Nature Neuroscience* 6, no. 2 (2003): 119–26; and the insular cortex, the region elected by A. D. Craig. A. D. Craig, *How Do You Feel? An Interoceptive Moment with Your Neurobiological Self* (Princeton, N.J.: Princeton University Press, 2015).

6. Although the essence of consciousness is mental and thus available only to the subject who is conscious, there is a long tradition of addressing consciousness from a perspective of behavior, from the outside in, as it were. Clinicians who work in emergency rooms, operating rooms, or intensive care units are trained in this external perspective and are ready to presume the presence or absence of consciousness on the basis of a silent observation or of a conversation with the patient if such conversation is possible. As a neurologist, I was trained to do this.

What does the clinician look for? Wakefulness, attentiveness, emotive animation, and purposeful gestures are helpful telltale signs of consciousness. Unconscious patients, as in cases of coma, are not awake, not attentive, not emotive, and the gestures they make, if any, are not meaningful relative to the environment. But the conclusions you may draw on this scenario are complicated by conditions in which consciousness is likely to be impaired, such as persistent vegetative states, but the person alternates sleep periods with awake periods. The problem of assuming the presence or absence of consciousness from external manifestations can get especially complicated in a condition known as locked-in syndrome. Here, consciousness is in fact maintained, but the patients are almost completely immobile, and it is easy to miss the subtle movements such patients can make, which largely consist of blinking and of limited eye movements. The clinical art has been refined to a point of reasonable safety, but still the only guaranteed way of establishing that someone is conscious is by having the person give direct testimony of a normal mental state. Clinicians are fond of declaring, after three questions relating to (a) the person's identity, (b) the place the person is in, and (c) the approximate date, that the person is or is not conscious. That is not comparable to knowing, directly and without equivocation, whether a person has a working, conscious mind.

There is a large literature on the neurological conditions that cause impairments of consciousness or that may appear to cause such impairments but in fact do not, such as locked-in syndrome. There is also an abundant literature on anesthesia and how the administration of varied chemical compounds disrupts mental experience reversibly. Both literatures provide important clues relative to the neural foundations of consciousness. It is fair to say, however, that the specific brain damage that causes coma or the chemical molecules responsible for anesthesia are blunt instruments that do not allow us to divine the neurobiological processes responsible for mental experience. Several anesthetics have the power to suspend the early process of sensing and responding that we find in bacteria, or in plants for that matter. Anesthetics freeze sensing and responding up and down the several branches of life. They do not suspend consciousness directly, but they do block processes on which mental states, feelings, and the perspectival stance depend. See Parvizi and Damasio, "Consciousness and the Brainstem"; Josef Parvizi and Antonio Damasio, "Neuroanatomical Correlates of Brainstem Coma," *Brain* 126, no. 7 (2003): 1524–36; Antonio Damasio and Kaspar Meyer, "Consciousness: An Overview of the Phenomenon and of Its Possible Neural Basis," in *The Neurology of Consciousness,* eds. Steven Laureys and Giulio Tononi (Burlington, Mass.: Elsevier, 2009), 3–14.

7. Eric D. Brenner, Rainer Stahlberg, Stefano Mancuso, Jorge Vivanco, František Baluška, and Elizabeth Van Volkenburgh, "Plant Neurobiology: An Integrated View of Plant Signaling," *Trends in Plant Science* 11, no. 8 (2006): 413–19; Lauren A. E. Erland, Christina E. Turi, and Praveen K. Saxena, "Serotonin: An Ancient Molecule and an Important Regulator of Plant Processes," *Biotechnology Advances* (2016); Jin Cao, Ian B. Cole, and Susan J. Murch, "Neurotransmitters, Neuroregulators, and Neurotoxins in the Life of Plants," *Canadian Journal of Plant Science* 86, no. 4 (2006): 1183–88; Nicolas Bouché and Hillel Fromm, "GABA in Plants: Just a Metabolite?," *Trends in Plant Science* 9, no. 3 (2004): 110–15.

This is the reason I differ in part from Arthur S. Reber's conclusions in "Caterpillars, Consciousness, and the Origins of Mind," *Animal Sentience* 1, no. 11 (2016). Single-celled organisms sense and respond, abilities that are fundamental for the later development of mind, feeling, and subjectivity, but they should not be regarded as mindful, feeling, and conscious.

8. Few authors have included feeling in a conception of consciousness, let alone conceived of consciousness from the point of view of

affect. Besides Jaak Panksepp and A. Craig, I encountered another exception in the work of Michel Cabanac; see Michel Cabanac, "On the Origin of Consciousness, a Postulate and Its Corollary," *Neuroscience and Biobehavioral Reviews* 20, no. 1 (1996): 33–40.

9. David J. Chalmers, "How Can We Construct a Science of Consciousness?," in *The Cognitive Neurosciences III*, ed. Michael S. Gazzaniga (Cambridge, Mass.: MIT Press, 2004), 1111–19; David J. Chalmers, *The Conscious Mind: In Search of a Fundamental Theory* (Oxford: Oxford University Press, 1996); David J. Chalmers, "Facing Up to the Problem of Consciousness," *Journal of Consciousness Studies* 2, no. 3 (1995): 200–219.

10 ON CULTURES

1. Charles Darwin, *On the Origin of Species* (New York: Penguin Classics, 2009); William James, *Principles of Psychology* (Hardpress, 2013); Sigmund Freud, *The Basic Writings of Sigmund Freud* (New York: Modern Library, 1995); Émile Durkheim, *The Elementary Forms of Religious Life* (New York: Free Press, 1995).

2. The idea that some aspects of cultures have biological origins remains controversial. The legacy of misguided incursions of biology in sociopolitical affairs is a reluctance from disciplines in the humanities and social sciences to admit biological findings in their midst. There is also a justified distaste for accounts of mental and social phenomena that reduce them in their entirety to biology and suffer from scientific triumphalism to boot. This is part of the two-cultures split of C. P. Snow legend. It was a problem half a century ago and regretably it remains a problem.

3. Edward O. Wilson, *Sociobiology* (Cambridge, Mass.: Harvard University Press, 1975). Sociobiology and its leader, E. O. Wilson, were not well received. See Richard C. Lewontin, *Biology as Ideology: The*

Doctrine of DNA (New York: HarperPerennial, 1991), for a critical perspective on sociobiology. Curiously, Wilson's position on affect was compatible with mine, as his subsequent work has continued to show. See E. O. Wilson, *Consilience* (New York: Knopf, 1998). See also William H. Durham, *Coevolution: Genes, Culture and Human Diversity* (Palo Alto, Calif.: Stanford University Press, 1991), as an example of the compatibility of biology and cultural processes.

4. Parsons, "Social Systems and the Evolution of Action Theory"; Parsons, "Evolutionary Universals in Society."

5. It is reasonable to think that beyond the processes that maintain chemical stability—the natural tendency of all matter to remain in the most stable conformations while less stable conformations vanish—there would be an additional process capable of leading a molecule to create another one like itself.

6. The degree of male violence is correlated with certain physical characteristics, which can be subsumed by the term "formidability." See Aaron Sell, John Tooby, and Leda Cosmides, "Formidability and the Logic of Human Anger," *Proceedings of the National Academy of Sciences* 106, no. 35 (2009): 15073–78.

7. Richard L. Velkley, *Being After Rousseau: Philosophy and Culture in Question* (Chicago: University of Chicago Press, 2002). Originally in Samuel Pufendorf and Friedrich Knoch, *Samuelis Pufendorfii Eris Scandica: Qua adversus libros De jure naturali et gentium objecta diluuntur* (Frankfurt-am-Main: Sumptibus Friderici Knochii, 1686).

8. The literature consulted for this section includes William James, *The Varieties of Religious Experience* (New York: Penguin Classics, 1983); Charles Taylor, *Varieties of Religion Today: William James Revisited* (Cambridge, Mass.: Harvard University Press, 2002); David Hume, *Dialogues Concerning Natural Religion and the Natural History of Religion* (New York: Oxford University Press, 2008); John R.

Bowen, *Religions in Practice: An Approach to the Anthropology of Religion* (Boston: Pearson, 2014); Walter Burkert, *Creation of the Sacred: Tracks of Biology in Early Religions* (Cambridge, Mass.: Harvard University Press, 1996); Durkheim, *Elementary Forms of Religious Life;* John R. Hinnells, ed., *The Penguin Handbook of the World's Living Religions* (London: Penguin Books, 2010); Claude Lévi-Strauss, *L'anthropologie face aux problèmes du monde moderne* (Paris: Seuil, 2011); Scott Atran, *In Gods We Trust: The Evolutionary Landscape of Religion* (New York: Oxford University Press, 2002).

9. Martha C. Nussbaum, *Political Emotions: Why Love Matters for Justice* (Cambridge, Mass.: Belknap Press of Harvard University Press, 2013); Jonathan Haidt, *The Righteous Mind: Why Good People Are Divided by Politics and Religion* (New York: Pantheon Books, 2012); Steven W. Anderson, Antoine Bechara, Hanna Damasio, Daniel Tranel, and Antonio Damasio, "Impairment of Social and Moral Behavior Related to Early Damage in Human Prefrontal Cortex," *Nature Neuroscience* 2 (1999): 1032–37; Joshua D. Greene, R. Brian Sommerville, Leigh E. Nystrom, John M. Darley, and Jonathan D. Cohen, "An fMRI Investigation of Emotional Engagement in Moral Judgment," *Science* 293, no. 5537 (2001): 2105–8; Mark Johnson, *Morality for Humans: Ethical Understanding from the Perspective of Cognitive Science* (University of Chicago Press, 2014); L. Young, Antoine Bechara, Daniel Tranel, Hanna Damasio, M. Hauser, and Antonio Damasio, "Damage to Ventromedial Prefrontal Cortex Impairs Judgment of Harmful Intent," *Neuron* 65, no. 6 (2010): 845–51.

10. Cyprian Broodbank, *The Making of the Middle Sea: A History of the Mediterranean from the Beginning to the Emergence of the Classical World* (London: Thames & Hudson, 2015); Malcolm Wiener, "The Interaction of Climate Change and Agency in the Collapse of Civilizations ca. 2300–2000 BC," *Radiocarbon* 56, no. 4 (2014): S1–S16; Malcolm Wiener, "Causes of Complex Systems Collapse at the End of the

Bronze Age," in *"Sea Peoples" Up-to-Date*, 43–74, Austrian Academy of Sciences (2014).

11. Karl Marx, *Critique of Hegel's "Philosophy of Right"* (New York: Cambridge University Press, 1970). As noted earlier the ideas of social scientists such as Bourdieu, Touraine, and Foucault also lend themselves to translation in biological terms.

12. Assal Habibi and Antonio Damasio, "Music, Feelings, and the Human Brain," *Psychomusicology: Music, Mind, and Brain* 24, no. 1 (2014): 92; Matthew Sachs, Antonio Damasio, and Assal Habibi, "The Pleasures of Sad Music: A Systematic Review," *Frontiers in Human Neuroscience* 9, no. 404 (2015): 1–12, doi:10.3389/fnhum.2015.00404.

13. From Antonio Damasio, "Suoni, significati affettivi e esperienze musicali," *Musica Domani*, 5–8, no. 176 (2017).

14. Sebastian Kirschner and Michael Tomasello, "Joint Music Making Promotes Prosocial Behavior in 4-Year-Old Children," *Evolution and Human Behavior* 31, no. 5 (2010): 354–64.

15. Panksepp, "Cross-Species Affective Neuroscience Decoding of the Primal Affective Experiences of Humans and Related Animals"; Henning et al., "A Role for mu-Opioids in Mediating the Positive Effects of Gratitude."

16. The contradictions posed by cutting, anorexia, and morbid obesity are simpler to address. It is a fact that people can indulge in the cutting of their skin, a practice that qualifies as cultural because it can spread by imitation and has seemingly random distribution. It is possible that the best explanation for these phenomena concerns the pathological circumstances of the affected individuals made worse by an equally pathological cultural context. The same applies to online communities of so-called gainers, individuals who gather and encourage each other to consume large amounts of food with the purpose of gaining weight, watch the results in each other, and engage in sex. To some extent both examples qualify for an old-fashioned diagnosis: masochism. The practice of masochism does produce pleasure, a situation that corresponds to an upregulation of homeostasis. It so happens

that the future, and ultimate costs of upregulation outweigh the gains, a physiological scenario not far from that of substance addictions. Pleasures give way to dependences and suffering. It is unlikely that such bizarre practices will be incorporated in biological evolution or be selected culturally beyond small groups. That the practices and groups even exist today testifies to the risks of fringe Internet communities.

17. Talita Prado Simão, Sílvia Caldeira, and Emilia Campos de Carvalho, "The Effect of Prayer on Patients' Health: Systematic Literature Review," *Religions* 7, no. 1 (2016): 11; Samuel R. Weber and Kenneth I. Pargament, "The Role of Religion and Spirituality in Mental Health," *Current Opinion in Psychiatry* 27, no. 5 (2014): 358–63; Neal Krause, "Gratitude Toward God, Stress, and Health in Late Life," *Research on Aging* 28, no. 2 (2006): 163–83.

18. Kirschner and Tomasello, "Joint Music Making Promotes Prosocial Behavior." Cited earlier.

19. Jason E. Lewis and Sonia Harmand, "An Earlier Origin for Stone Tool Making: Implications for Cognitive Evolution and the Transition to *Homo*," *Philosophical Transactions of the Royal Society B* 371, no. 1698 (2016): 20150233.

20. Robin I. M. Dunbar and John A. J. Gowlett, "Fireside Chat: The Impact of Fire on Hominin Socioecology," *Lucy to Language: The Benchmark Papers,* ed. Robin I. M. Dunbar, Clive Gamble, and John A. J. Gowlett (New York: Oxford University Press, 2014), 277–96.

21. Polly W. Wiessner, "Embers of Society: Firelight Talk Among the Ju/'hoansi Bushmen," *Proceedings of the National Academy of Sciences* 111, no. 39 (2014): 14027–35.

11 MEDICINE, IMMORTALITY, AND ALGORITHMS

1. Jennifer A. Doudna and Emmanuelle Charpentier, "The New Frontier of Genome Engineering with CRISPR-Cas9," *Science* 346, no. 6213 (2014): 1258096.

2. Pedro Domingos, *The Master Algorithm: How the Quest for the Ultimate Learning Machine Will Remake Our World* (New York: Basic Books, 2015).

3. Krishna V. Shenoy and Jose M. Carmena, "Combining Decoder Design and Neural Adaptation in Brain-Machine Interfaces," *Neuron* 84, no. 4 (2014): 665–80, doi:10.1016/j.neuron.2014.08.038; Johan Wessberg, Christopher R. Stambaugh, Jerald D. Kralik, Pamela D. Beck, Mark Laubach, John K. Chapin, Jung Kim, S. James Biggs, Mandayam A. Srinivasan, and Miguel A. Nicolelis, "Real-Time Prediction of Hand Trajectory by Ensembles of Cortical Neurons in Primates," *Nature* 408, no. 6810 (2000): 361–65; Ujwal Chaudhary et al., "Brain-Computer Interface-Based Communication in the Completely Locked-In State," *PLoS Biology* 15, no. 1 (2017): e1002593, doi:10.1371/journal.pbio.1002593; Jennifer Collinger, Brian Wodlinger, John E. Downey, Wei Wang, Elizabeth C. Tyler-Kabara, Douglas J. Weber, Angus J. McMorland, Meel Velliste, Michael L. Boninger, and Andrew B. Schwartz, "High-Performance Neuroprosthetic Control by an Individual with Tetraplegia," *Lancet* 381, no. 9866 (2013): 557–64, doi:10.1016/S0140-6736(12)61816-9.

4. Ray Kurzweil, *The Singularity Is Near: When Humans Transcend Biology* (New York: Penguin, 2005); Luc Ferry, *La révolution transhumaniste: Comment la technomédecine et l'uberisation du monde vont bouleverser nos vies* (Paris: Plon, 2016).

5. Yuval Noah Harari, *Homo Deus: A Brief History of Tomorrow* (Oxford: Signal Books, 2016).

6. Nick Bostrom, *Superintelligence: Paths, Dangers, Strategies* (Oxford: Oxford University Press, 2014).

7. Margalit, *Ethics of Memory*.

8. Aldous Huxley, *Brave New World* (1932; London: Vintage, 1998).

9. George Zarkadakis, *In Our Own Image: Savior or Destroyer? The History and Future of Artificial Intelligence* (New York: Pegasus Books, 2015).

10. W. Grey Walter, "An Imitation of Life," *Scientific American* 182, no. 5 (1950): 42–45.

12 ON THE HUMAN CONDITION NOW

1. Epicurus and Bertrand Russell would have been pleased to know that their philosophical concerns for human happiness have not been forgotten. Epicurus, *The Epicurus Reader*, eds. B. Inwood and L. P. Gerson (Indianapolis: Hackett, 1994); Bertrand Russell, *The Conquest of Happiness* (New York: Liveright, 1930); Daniel Kahneman, "Objective Happiness," in *Well-Being: Foundations of Hedonic Psychology*, eds. Daniel Kahneman, Edward Diener, and Norbert Schwarz (New York: Russell Sage Foundation, 1999); Amartya Sen, "The Economics of Happiness and Capability," in *Capabilities and Happiness*, eds. Luigino Bruni, Flavio Comim, and Maurizio Pugno (New York: Oxford University Press, 2008); Richard Davidson and Brianna S. Shuyler, "Neuroscience of Happiness," in *World Happiness Report 2015*, eds. John F. Helliwell, Richard Layard, and Jeffrey Sachs (New York: Sustainable Development Solutions Network, 2015).

2. Neil Postman, *Amusing Ourselves to Death: Public Discourse in the Age of Show Business* (New York: Penguin, 2006). See also Robert D. Putnam, *Our Kids* (New York: Simon & Schuster, 2015).

3. Jonas T. Kaplan, Sarah I. Gimbel, and Sam Harris, "Neural Correlates of Maintaining One's Political Beliefs in the Face of Counterevidence," *Nature Scientific Reports* 6 (2016).

4. Sherry Turkle, *Alone Together: Why We Expect More from Technology and Less from Each Other* (New York: Basic Books, 2011); Alain Touraine, *Pourrons-nous vivre ensemble?* (Paris: Fayard, 1997).

5. Manuel Castells, *Communication Power* (New York: Cambridge University Press, 2009); Manuel Castells, *Networks of Outrage and Hope: Social Movements in the Internet Age* (New York: John Wiley & Sons, 2015).

6. Amartya Sen, "The Economics of Happiness and Capability"; Onora O'Neill, *Justice Across Boundaries: Whose Obligations?* (Cambridge: Cambridge University Press, 2016); Nussbaum, *Political Emotions;* Peter Singer, *The Expanding Circle: Ethics, Evolution, and Moral Progress* (Princeton, N.J.: Princeton University Press, 2011); Steven Pinker, *The Better Angels of Our Nature: Why Violence Has Declined* (New York: Penguin Books, 2011).

7. See Haidt, *Righteous Mind.*

8. Sigmund Freud, *Civilization and Its Discontents: The Standard Edition* (New York: W. W. Norton, 2010).

9. Albert Einstein and Sigmund Freud, *Why War? The Correspondence Between Albert Einstein and Sigmund Freud,* trans. Fritz Moellenhoff and Anna Moellenhoff (Chicago: Chicago Institute for Psychoanalysis, 1933).

10. Janet L. Lauritsen, Karen Heimer, and James P. Lynch, "Trends in the Gender Gap in Violent Offending: New Evidence from the National Crime Victimization Survey," *Criminology* 47, no. 2 (2009): 361–99; Richard Wrangham and Dale Peterson, *Demonic Males: Apes and the Origins of Human Violence* (Boston and New York: Houghton Mifflin Company, 1996); Sell, Tooby, and Cosmides, "Formidability and the Logic of Human Anger."

11. Zivin, Hsiang, and Neidell, "Temperature and Human Capital in the Short- and Long-Run"; Butke and Sheridan, "Analysis of the Relationship Between Weather and Aggressive Crime in Cleveland, Ohio."

12. Harari, *Homo Deus;* Bostrom, *Superintelligence.*

13. Parsons, "Evolutionary Universals in Society."

14. Thomas Hobbes, *Leviathan* (New York: A&C Black, 2006); Jean-Jacques Rousseau, *A Discourse on Inequality* (New York: Penguin, 1984).

15. John Gray, *Straw Dogs: Thoughts on Humans and Other Animals* (New York: Farrar, Straus and Giroux, 2002); John Gray, *False Dawn: The Delusions of Global Capitalism* (London: Granta, 2009); John Gray, *The Silence of Animals: On Progress and Other Modern Myths* (New York: Farrar, Straus and Giroux, 2013).

16. Max Horkheimer and Theodor W. Adorno, *Dialectic of Enlightenment: Philosophical Fragments* (Stanford, Calif.: Stanford University Press, 2002).

17. "Burden" is an especially appropriate term for a good part of the effects of consciousness. We owe the usage to George Soros, *The Age of Fallibility: Consequences of the War on Terror* (New York: Public Affairs, 2006).

18. On this issue read a valuable monograph by David Sloan Wilson, *Does Altruism Exist? Culture, Genes, and the Welfare of Others* (New Haven, Conn.: Yale University Press, 2015).

19. Verdi wrote *Falstaff* in 1893. A decade earlier and not terribly far, Richard Wagner, never having managed to separate love from death, was still consumed by pagan mayhem. His closest approximation to a sunny version of the human condition was the redemptive *Parsifal*.

20. Paul Bloom's qualification of empathy is relevant in this regard. Paul Bloom, *Against Empathy: The Case for Rational Compassion* (New York: HarperCollins, 2016).

13 THE STRANGE ORDER OF THINGS

1. D'Arcy Thompson, "On Growth and Form," in *On Growth and Form* (Cambridge, U.K.: Cambridge University Press, 1942).

2. Howard Gardner, *Truth, Beauty, and Goodness Reframed: Educating for the Virtues in the Twenty-First Century* (New York: Basic Books, 2011); Mary Helen Immordino-Yang, *Emotions, Learning, and the Brain: Exploring the Educational Implications of Affective Neuroscience* (New York: W. W. Norton, 2015); Wilson, *Does Altruism Exist?*; Mark Johnson, cited earlier.

3. Colin Klein and Andrew B. Barron, "Insects Have the Capacity for Subjective Experience," *Animal Sentience* (2016): 100; Peter Godfrey-Smith, *Other Minds: The Octopus, the Sea, and the Deep Origins of Consciousness* (New York: Farrar, Straus and Giroux, 2016). On the issue of non-human behavioral and cognitive abilities, I am in firm agreement with the position of Frans De Waal, Jaak Panksepp, and a growing number of biologists and cognitive scientists. As noted elsewhere the exceptional position of humans does not require diminishing the abilities of other animals. On the other hand, while allowing plenty of intelligent behavior to the earliest of living species, I hypothesize that well-adapted intelligence does not signify consciousness, a point on which Arthur Reber and I differ. The journal *Animal Sentience*, edited by Steven Harnad, is a new and excellent forum for scholarship on these issues.

4. In a recent essay on the mind and body problem, Siri Hustvedt voices the same view. Siri Hustvedt, *A Woman Looking at Men Looking at Women: Essays on Art, Sex, and the Mind* (New York: Simon & Schuster, 2016).

5. Seth, "Interoceptive Inference, Emotion, and the Embodied Self."

Machelska, Halina, 271n2
Macho, Alberto P., 275n9
Mahabharata, 177
Malthus, Thomas, 165
mammals
  affective apparatus of, 18
  drives and motivations of, 114,
    270–71n11
  emotive response of, 110–11
  enteric nervous systems of, 135
  layered feeling states of, 116
  mapmaking minds of, 78
  mutual grooming of, 183
  sociality of, 269–71n10
Man, Kingson, 246, 262n6, 275n10
Mancuso, Stefano, 259n5
manic states, 106
mapping and image-making, 61–63,
    74–83, 99, 121, 123, 165, 242–44
  consciousness and, 90, 149–56
  expansion of minds and, 84–98, 168
  of external worlds, 76–77, 79–80,
    85–89, 126, 127–28, 155–56, 186
  homeostasis and, 80–81
  integration of experience in, 153–56,
    238
  of internal states, 76–77, 80–83,
    85–87, 89, 126, 186
  meaning-making and symbolic
    thought in, 89–90, 243, 264n4
  memory and, 71, 91, 93–97, 186–87
  narrative and storytelling in, 91–92,
    96–97, 159, 264n6
  organism-specific contexts of, 159
  production of feelings and, 100–104,
    123–24
  sensory integration in, 87–92
  subjectivity and, 149–53, 281n1,
    282n3
Margalit, Avishai, 265n8, 290n7
Margulis, Lynn, 55, 257n2
Marx, Karl, 178, 288n11
masochism, 106, 184, 288n16
Materne, Eva, 252n7
Mather, Mara, 280n30
Maturana, Humberto R., 39, 254–55n11
Maupertuis, Pierre Louis Moreau de,
    35–36, 252n9, 253n3

Mayer, Emeran, 279n25
McComb, Karen, 269n10
McCulloch, Warren S., 240, 259–60n11
McDermott, Rose, 246
McEwen, Bruce S., 257n9, 266–67n7
McFall-Ngai, Margaret, 54, 257n1
McKenna, Charles, 246
McKenzie, Ian A., 278n23
McKinley, Michael J., 277n18
McMahon, Stephen B., 271–73nn2–3
media. *See* news media
medial cortical regions, 155, 215
medicine, 4–5, 194–98
  artificial intelligence and diagnostics
    in, 197
  genetic manipulation in, 195–96
  pain management and, 209–10
  robotics in, 196–98
  traditional tools and technology of,
    194–95
meditation, 152
Medzhitov, Ruslan, 256n3
memory, 62–63, 71, 165, 168, 243,
    265n2
  emotive response and, 112
  hippocampus and, 94–95
  image-making and, 71, 91, 93–97,
    186–87
  imagination and, 96–97, 186–87, 243
  of motor-related activities, 95–96
  prediction of the future and, 97, 142
  reasoning and, 96, 187
  transformation of recollected feelings
    in, 140–42
mental states. *See* feelings and emotions
The Merry Wives of Windsor
    (Shakespeare), 231
metabolism, 33–34, 50, 77, 80–81
  chemical interoception in, 58
  role of genes in, 39–42
  See also homeostasis
metabolism-first view of origins, 38–42
metazoans, 54–55
Meyer, Kaspar, 262–63n6, 282–84n6
Meyer, Richard A., 278–79n24
Miller, Bennett, 246
Miller, Gregory E., 266–67n7
Miller, Stanley L., 38, 254n10

Salzet, Beatrice, 271n22
Sampson, Timothy, 280n28
Santoni, Giorgio, 271–72n3
Sche, Stefan, 252n7
Schrödinger, Erwin, 7, 40–41, 255n12
Schwann cells, 132
Schwartz, Andrew B., 290n3
Schwartz, James H., 259n4
Schwartz, Norbert, 291n1
science fiction, 206–7
sciences, 179, 182–83
second brain, 134–35
Segerstrom, Suzanne C., 266–67n7
self-mutilation, 184, 288n16
Sell, Aaron, 286n6, 292n10
Semple, Stuart, 269n10
Sen, Amartya, 218, 291n1, 292n6
sensory information, 95–96
    in bacteria and simple organisms,
        72–74, 76, 119, 121–24, 235–36,
        238–39, 260n2, 271–72n3,
        273–75nn5–6, 284n7
    drug addiction and, 209–10
    emotive responses and, 85–87,
        99–100, 108–15
    image-making and, 149–52, 155–56
    integration of, 87–92, 154–58
    memory and, 93–97
    pain and, 118–20, 136–38, 271–72n3,
        273–74n5
    perception of, 127–28
    peripheral nervous system and,
        128–31, 136–38, 240–42
    somatosensory system and, 150–51
    subcortical control apparatus of,
        110–13, 238–39, 268–69n9
    technological enhancement of, 197
    telesenses, 128
sensory portals, 79–83, 85–87, 126, 150
serotonin, 51, 103, 136
Seth, Anil K., 259n7, 294n5
Shackman, Alexander J., 276n13
Shakespeare, William, 177, 231
Shannon, Claude, 240
Shapiro, Robert E., 277n18
Shenoy, Krishna V., 290n3
Sheridan, Scott C., 255n1
Shuyler, Brianna S., 291n1
Simmons, W. Kyle, 261n3

simple organisms. *See* bacteria and
    simple organisms; cellular life
Singer, Peter, 218, 292n6
Sisyphus, 224
slavery, 225
smell. *See* olfaction
Smirnov, Igor, 277n17
Smith, John Maynard, 254n11
Snow, C. P., 285n2
social Darwinism, 165–66
sociality, 3–5, 15, 190, 234–37
    of bacteria and single-celled
        organisms, 54, 167–69, 234–36,
        269–71n10
    of drives, motivations, and emotions,
        113–15, 269–71nn10–11
    of insects, 22–24, 234
social media, 216–17
sociobiology, 166, 285–86n3
Sokabe, Takaaki, 271–73n3
Soll, Jacob, 246
Solms, Mark, 249–50n2
somatosensory system, 150–51, 155
Soros, George, 293n17
Spencer, Herbert, 165
Sperandio, Vanessa, 251n6, 257–58n3
spinal cord, 62
Spinoza, Baruch de, 35–36, 40–41, 240,
    253n4, 276n11
Spitzer, Jan, 254–55n11
spontaneous feelings, 99, 107, 109
sports, 174
Stefano, George B., 271n2
Stein, Christoph, 271n2
storytelling, 7, 91–92, 96–97, 159,
    181–82, 187, 264n6
stress, 117, 183, 266–67n7
Strick, Peter L., 277n17
subcortical nuclei, 110–11, 277nn14–15
subjectivity, 62, 72–73, 83, 91, 120, 186,
    281n1, 282n3
    consciousness and, 143–44, 147–53,
        160, 228–33, 237–39
    cultural invention and, 158–59
    duality of, 126, 276n11
    evolutionary ancestors to, 110–13,
        238–39
    feelingness and, 152–53
    image-making and, 149–52, 168

ABOUT THE AUTHOR

Antonio Damasio is University Professor; David Dornsife Professor of Neuroscience, Psychology and Philosophy; and director of the Brain and Creativity Institute at the University of Southern California, Los Angeles.

Trained as both neurologist and neuroscientist, Damasio has made seminal contributions to the understanding of brain processes underlying emotions, feelings, and consciousness. His work on the role of affect in decision-making has made a major impact in neuroscience, psychology, and philosophy. He is the author of numerous scientific articles and has been named "Highly Cited Researcher" by the Institute for Scientific Information, and is regarded as one of the most eminent psychologists of the modern era.

Damasio is a member of the National Academy of Medicine and a fellow of the American Academy of Arts and Sciences, the Bavarian Academy of Sciences, and the European Academy of Sciences and Arts. He has received numerous prizes, among them the Grawemeyer Award (2014) and the Honda Prize (2010), the Asturias Prize in Science and Technology (2005), and the Nonino (2003), Signoret (2004), and Pessoa (1992) prizes.

He holds honorary doctorates from several leading universities, some shared with his wife, Hanna, among them the École Polytechnique Fédérale de Lausanne (EPFL), 2011, and the Sorbonne (Université Paris Descartes), 2015.

Damasio has discussed his research and ideas in several books, among them *Descartes' Error, The Feeling of What Happens, Looking for Spinoza* and *Self Comes to Mind,* which are translated and taught in universities worldwide.

For more information, go to the Brain and Creativity Institute Web site at www.usc.edu/bci/.

A NOTE ON THE TYPE

The text of this book was set in a typeface called Aldus, designed by the celebrated typographer Hermann Zapf in 1952–1953. Based on the classical proportion of the popular Palatino type family, Aldus was originally adapted for Linotype composition as a slightly lighter version that would read better in smaller sizes.

Hermann Zapf was born in Nuremberg, Germany, in 1918. He has created many other well-known typefaces including Comenius, Hunt Roman, Marconi, Melior, Michelangelo, Optima, Saphir, Sistina, Zapf Book, and Zapf Chancery.

Typeset by North Market Street Graphics, Lancaster, Pennsylvania

Printed and bound by Berryville Graphics, Berryville, Virginia

Designed by Maggie Hinders